BACTERIOLOGY RESEARCH DEVELOPMENTS

ENTEROCOCCUS FAECALIS

MOLECULAR CHARACTERISTICS, ROLE IN NOSOCOMIAL INFECTIONS AND ANTIBACTERIAL EFFECTS

BACTERIOLOGY RESEARCH DEVELOPMENTS

Additional books in this series can be found on Nova's website
under the Series tab.

Additional e-books in this series can be found on Nova's website
under the e-book tab.

BACTERIOLOGY RESEARCH DEVELOPMENTS

ENTEROCOCCUS FAECALIS

MOLECULAR CHARACTERISTICS, ROLE IN NOSOCOMIAL INFECTIONS AND ANTIBACTERIAL EFFECTS

HENRY L. MACK
EDITOR

Copyright © 2014 by Nova Science Publishers, Inc.

All rights reserved. No part of this book may be reproduced, stored in a retrieval system or transmitted in any form or by any means: electronic, electrostatic, magnetic, tape, mechanical photocopying, recording or otherwise without the written permission of the Publisher.

For permission to use material from this book please contact us:
Telephone 631-231-7269; Fax 631-231-8175
Web Site: http://www.novapublishers.com

NOTICE TO THE READER

The Publisher has taken reasonable care in the preparation of this book, but makes no expressed or implied warranty of any kind and assumes no responsibility for any errors or omissions. No liability is assumed for incidental or consequential damages in connection with or arising out of information contained in this book. The Publisher shall not be liable for any special, consequential, or exemplary damages resulting, in whole or in part, from the readers' use of, or reliance upon, this material. Any parts of this book based on government reports are so indicated and copyright is claimed for those parts to the extent applicable to compilations of such works.

Independent verification should be sought for any data, advice or recommendations contained in this book. In addition, no responsibility is assumed by the publisher for any injury and/or damage to persons or property arising from any methods, products, instructions, ideas or otherwise contained in this publication.

This publication is designed to provide accurate and authoritative information with regard to the subject matter covered herein. It is sold with the clear understanding that the Publisher is not engaged in rendering legal or any other professional services. If legal or any other expert assistance is required, the services of a competent person should be sought. FROM A DECLARATION OF PARTICIPANTS JOINTLY ADOPTED BY A COMMITTEE OF THE AMERICAN BAR ASSOCIATION AND A COMMITTEE OF PUBLISHERS.

Additional color graphics may be available in the e-book version of this book.

Library of Congress Cataloging-in-Publication Data

ISBN: 978-1-63321-049-3

LCCN: 2014939984

Published by Nova Science Publishers, Inc. † New York

CONTENTS

Preface		vii
Chapter 1	Virulence Factors of Enterococcus faecalis: The Promoters for Their Pathogenicity *Joana Barbosa, Sandra Borges and Paula Teixeira*	1
Chapter 2	Naturally-Derived Molecules As a Strategy for Countering *E. faecalis* Infection *David M. Pereira*	19
Chapter 3	*Enterococcus faecalis:* Role in Nosocomial Infection, Resistance Traits and Molecular Epidemiology *Juliana Caierão*	29
Chapter 4	Promiscuity, Pheromones and Pathogenicity: Why All Enterococci Are Not Created Equal *Elise Pelzer, Irani Rathnayake and Flavia Huygens*	69
Chapter 5	High-Level Gentamicin Resistance in *Enterococcus faecalis:* Molecular Characteristics and Relevance in Severe Infections *Mónica Sparo and G. Delpech*	93
Chapter 6	*Enterococcus faecalis* in Endodontics *María Gabriela Pacios and María Elena López*	109
Chapter 7	Molecular Characterization of Natural Dairy Isolates of *Enterococcus faecalis* and Evaluation of Their Antimicrobial Potential *Katarina Veljović, Amarela Terzić-Vidojević, Maja Tolinački, Sanja Mihajlović, Goran Vukotić, Natasa Golić and Milan Kojić*	123
Chapter 8	*Enterococcus faecalis* in Dental Infections: Virulence Factors, Molecular Characteristics, Antibacterial and Anti-Infective Techniques *Nurit Beyth, Ronit Poraduso-Cohen and Ronen Hazan*	137
Index		165

PREFACE

Enterococcus faecalis is a gram-positive, coccus shaped, lactic acid bacterium, with demonstrated ubiquity across multiple anatomical sites. Enterococcus faecalis isolates have been isolated from clinical samples as the etiological agent in patients with overt infections, and from body sites previously thought to be sterile but absent of signs and symptoms of infection. E. faecalis is implicated in both human health and disease, recognized as a commensal, a probiotic and an opportunistic multiply resistant pathogen. E. faecalis has emerged as a key pathogen in nosocomial infections. Enterococcus faecalis is a commensal bacterium inhabiting the gastro-intestinal tract of humans. Interestingly, although it is not clear whether E. faecalis is part of the oral cavity microbiome, it is frequently recovered from root canal infections. Specifically, it is the major pathogen found in persistent infections associated with root canal treatment failure. Moreover, E. faecalis is one of the leading multidrug resistant nosocomial pathogens, causing infective endocarditis, and participating in urinary tract, wound, and device- device-related infections. This book discusses the molecular characteristics, its role in nonsocomial infections and the antibacterial effects of Enterococcus faecalis. It begins by discussing the virulence factors of enterococcus faecalis and concludes with E. faecalis in dental infections.

Chapter 1 – Enterococci are natural inhabitants of the intestinal microbiota of humans and animals. They can survive in many environments contaminated with human or animal faecal materials (such as water, soils receiving and any surfaces) as well as food products derived from animals. They are important causes of community-acquired and nosocomial infections, such as urinary tract infections, endocarditis and bacteraemia. Among enterococcal diseases, the majority are associated with *Enterococcus faecalis* species. Several factors are associated with a high risk of acquiring enterococcal infections, including antimicrobial resistance and expression of virulence determinants. The most known virulence factors are enterococcal surface protein, aggregation substance, cell wall adhesins, sex pheromones, production of extracellular superoxide, production of hydrolytic enzymes (such as gelatinase), secretion of cytolysin, production of biofilm, among others.

To cause infection, *E. faecalis* must have virulence traits which allow the infecting strains to colonize and invade host tissue and translocate through epithelial cells and evade the host's immune response.

This review explores the importance of the presence of virulence factors for understanding the potential pathogenic activity of *E. faecalis*.

Chapter 2 – *Enterococcus faecalis* is a gram-positive bacteria that, while a frequent gut commensal, is one of the leading causes of nosocomial infections, which comprise urinary tract infections, endocarditis, bacteremia and meningitis. An important clinical feature of this species is the resistance to a wide range of antimicrobial agents, as demonstrated in clinical, food and water isolates. It not only contains several natural antibiotic resistances, but it is also capable of acquiring new ones as a result of mutations or by acquisition of new genes. Thus, there is a continuous need to search for new drugs that may be used against *E. faecalis*. Some naturally occurring chemical compounds have played a central role in antibiotic drug discovery, with a very significant percentage of clinically proven drugs being derived from natural products. Recently, studies have been reporting that even commonly used herbs, fruits or vegetables, may contain molecules that could constitute potential new treatment against several bacterial infections, including multi-drug resistant bacteria. The present chapter focuses on the most recent published reports on naturally-derived antimicrobial molecules effective against *E. faecalis*. When available, the molecular mechanism of action will also be addressed.

Chapter 3 – Enterococci are recognized by their physiological versatility, which is responsible for the ubiquitous occurrence of these microorganisms. Because of this extraordinary capacity to survive under unfavorable conditions, they can persist in nosocomial environment for long periods, which may represent the source of exogenous enterococcal infections.

In humans, they compose genitourinary, oral and especially gastrointestinal microbiota. *Enterococcus faecalis* is, by far, the major species in both, colonized or infected patients. Although *Enterococcus faecium* is well-recognized by its resistance, *E. faecalis* consistently presents a more robust virulence arsenal. Its virulence has been defined as multifactorial, with participation of many different molecules and features, especially adhesins and the capacity of biofilm production.

Until the 70s, their pathogenic role had been neglected. However, since that period, they have been recognized as one of the leading cause of opportunistic infections in nosocomial setting, especially affecting immunosupressed, elderly or long-term hospitalized patients. The major clinical syndromes related to enterococci are bacteremia, endocarditis and urinary tract infections. Besides, they may be frequently associated to biliary, abdominal and wound infections.

The acceptance of this pathogenicity was coincident, and possibly related with the increase use of broad-spectrum antimicrobial agents, such as third-generation cephalosporins, to whom enterococci are intrinsically resistant. Indeed, this intrinsic resistance is extended to the majority of antimicrobials commonly used to treat gram-positive cocci infections. Therefore, this feature makes them much more adapted to the nosocomial environment than other bacterial genus.

Along with their intrinsic resistance characteristics, enterococci present an extraordinary capacity to acquire mobile genetic elements, carrying resistance genes to different classes of antimicrobials, including: chloramphenicol, tetracyclines, macrolides, glycopeptides and high levels of aminoglycoside. Therefore, efficient antimicrobials are scarce, leading to difficulties in treatment of enterococcal infections. The occurrence and dissemination of multidrug-resistant strains is well-recognized and represents a challenge to medical and infection control staff.

The most impactful phenotype is the Vancomycin-Resistant Enterococci (VRE), commonly related to multidrug-resistant isolates. Despite the introduction of new drugs active against VRE, resistance to them, including linezolid, tigecycline and daptomycin, have been described around the world. VRE genotype may be related to nine different *van* genes, located in mobile genetic elements. The most relevant genotype is, by far, *vanA*-VRE, which may present a clonal or heterogeneous dissemination, although the former seems to be more frequent and is associated with exogenous acquisition through healthcare staff. Long-term hospitalization, stay in intensive therapy unit and previous usage of antimicrobials have been associated to VRE acquisition.

VRE is endemic in many regions of the world and management of outbreaks requires strategies to avoid new cases and to reduce transmission rates, which includes isolation of infected or colonized patients. This management is difficult because, once established in determined nosocomial setting, VRE is hard to eradicate.

In conclusion, enterococci is a challenging opportunistic pathogen in nosocomial settings, especially because of its resistance traits, ability to survive for long periods in the environment, difficulty to eradicate and control its dissemination and the multifactorial virulence, which can cause life threatening infections in specific and severely ill patients.

Chapter 4 – *Enterococcus faecalis* is a Gram-positive, coccus shaped, lactic acid bacterium, with demonstrated ubiquity across multiple anatomical sites. *Enterococcus faecalis* isolates have been isolated from clinical samples as the etiological agent in patients with overt infections, and from body sites previously thought to be sterile but absent of signs and symptoms of infection.

E. faecalis is implicated in both human health and disease, recognized as a commensal, a probiotic and an opportunistic multiply resistant pathogen. *E. faecalis* has emerged as a key pathogen in nosocomial infections.

E. faecalis is well equipped to avert recognition by host cell immune mediators. Antigenic cell wall components including lipotechoic acids are concealed from immune detection by capsular polysaccharides produced by some strains. Thereby preventing complement activation, the pro-inflammatory response, opsonisation and phagocytosis. *E. faecalis* also produces a suite of enzymes including gelatinase and cytolysin, which aid in both virulence and host immune evasion. The ability of enterococci to form biofilms *in vivo* further increases virulence, whilst simultaneously preventing detection by host cells.

E. faecalis exhibits high levels of both intrinsic and acquired antimicrobial resistance. The mobility of the *E. faecalis* genome is a significant contributor to antimicrobial resistance, with this species also transferring resistance to other Gram-positive bacteria.

Whilst *E. faecalis* is of increasing concern in nosocomial infections, its role as a member of the endogenous microbiota cannot be underestimated. As a commensal and probiotic, *E. faecalis* plays an integral role in modulating the immune response, and in providing endogenous antimicrobial activity to enhance exclusion or inhibition of opportunistic pathogens in certain anatomical niches.

In this chapter the authors will review possible mediators of enterococcal transition from commensal microbe to opportunistic pathogen, considering isolates obtained from patients diagnosed with pathogenic infections and those obtained from asymptomatic patients.

Chapter 5 – The genus *Enterococcus* belongs to the indigenous gastrointestinal microbiota of humans and animals, and is present in food of animal origin as well as in vegetables. Whenever these bacteria cause invasive infections, their eradication is difficult.

This phenomenon is linked to the natural and acquired antimicrobial resistance of enterococci as well as the existence of cell components that can behave as virulence factors. Cytolysin production contributes with the severity of infectious diseases in animal models and in humans. It has been proven that 60% of *Enterococcus faecalis* strains isolated from different infections sites are cytolysin producers. Their presence is associated with a five-fold increase of death risk in bacteremic patients. Clinical and microbiological resolution of severe infections can be affected when they are caused by cytolysin-producers *E. faecalis* strains that display high-level gentamicin resistance (HLGR). Enterococci carry a wide variety of mobile genetic elements and they are regarded as a reservoir of acquired antimicrobial resistance genes for Gram-positive bacteria. Multiple antimicrobial resistance is common among enterococci and constitutes a relevant Public Health issue. In 1979, HLGR (MIC ≥ 500 µg/mL) in enterococci was reported for the first time. HLGR represents a significant therapeutic problem for human medicine, especially for patients with invasive infectious diseases that require bactericidal efficacy such as meningitis, endocarditis and osteomyelitis. The most frequent HLGR gene among enterococci is *aac (6´)-ie-aph (2´´)-Ia* that encodes AAC(6´)-APH(2´´), an enzyme with acetyl transferase and phospho transferase activities. This bifunctional gene confers resistance to aminoglycosides available for therapeutic use with the exception of streptomycin. As a consequence, the synergistic role (bactericidal effect) of aminoglycosides with cell wall-active agents such as ampicillin or vancomycin is precluded. Along the last decade, *E. faecalis*has emerged as a relevant health-care associated pathogen. An identical mechanism for HLGR has been reported for human, animal and food enterococcal strains. In addition, exchange of these resistance genes through horizontal transfer is feasible. Risk factors for the acquisition of infection with high-level gentamicin resistant enterococci have been identified: previous long-term antimicrobial treatment, number of prescribed antimicrobials, previous surgeries, perioperative antimicrobial prophylaxis, hospitalization term/ antimicrobial treatment, urinary catheterization and renal failure. Infections caused by high-level gentamicin resistant *E. faecalis* constitute a severe risk for patients with invasive conditions and long-term hospitalization. Clonal expansion and emergence of unique bacterial strains contribute to the significant enhancement of infectious diseases caused by high-level gentamicin resistant *E. faecalis*.

Chapter 6 – Enterococcus faecalis is a persistent agent that frequently causes infection of the tooth root canal and failure of endodontic treatments. The infection is hard to be treated. Chemo mechanical cleaning and shaping of the root canal can greatly reduce the number of bacteria. However, the use of intracanal medications to disinfect the root canal system has been advocated to enhance the success of the treatment.

Calcium hydroxide is widely used as an intracanal medicament in endodontic therapy. The in vitro calcium hydroxide action and its vehicles evaluated against Enterococcus faecalis showed that this bacterium had the higher inhibition zones with calcium hydroxide + p-monochlorophenol; calcium hydroxide + p-monochlorophenol-propylene glycol pastes and 2% chlorhexidine gluconate in comparison to other intracanal medicaments. The vehicle used to prepare the calcium hydroxide paste might contribute to its antibacterial action. Chlorhexidine gluconate gel used alone, and camphorated p-monochlorophenol and camphorated p-monochlorophenol- propylene glycol as vehicles of calcium hydroxide, could be recommended, in an antimicrobial sense.

Since chlorhexidine gluconate may be used in different forms, another experience demonstrated the best of them to evidence efficacy to eliminate the most resistant intracanal bacteria. Maxillary anterior human teeth were prepared, sterilized and infected with Enterococcus faecalis for 3 days. Specimens were filled with calcium hydroxide + distilled water, 2% chlorhexidine gel and calcium hydroxide + 2% chlorhexidine (aqueous solution) and incubated at 37ºC. At different times the dressings were removed and the teeth were immersed in Brain Heart Infusion broth. Enterococcus faecalis growth was evaluated by monitoring turbidity of the culture medium. Specimens were observed by scanning electron microscopy. Chlorhexidine gel was effective in eliminating Enterococcus faecalis at day 1. Calcium hydroxide + distilled water and calcium hydroxide + chlorhexidine aqueous solution showed no antimicrobial effect on Enterococcus faecalis. Scanning electron microscopy observations evidenced these results. Chlorhexidine gel was the only effective intracanal medicament against Enterococcus faecalis.

A common clinical problem in Endodontics which is the re-infection of the root canal by Enterococcus faecalis, and ultimately, the failure of the endodontic treatment, may be satisfactory solved by the use of 2% chlorhexidine gluconate gel.

Chapter 7 – Due to their ability to survive adverse conditions, enterococci are widespread in nature and can be found in milk, dairy products and human and animal gastrointestinal tracts. Still, the use of enterococci in food preparation is controversial, since they have traditionally been branded as indicators of faecal contamination and their role in food spoilage is well known. However, some enterococcal strains exhibit antimicrobial effects and have probiotic potential, contributing to the improvement of the general state of health. For that reason, the authors have analyzed natural isolates of *Enterococcus faecalis* originating from various dairy products manufactured in rural households located in the mountains of Serbia. Genotyping analysis of selected enterococci showed high diversity among the isolates. The antimicrobial activity of the isolates showed a great effect on the number of pathogenic and non-pathogenic strains, including *L. monocytogenes*, *L. innocua*, *E. coli*, *Pseudomonas sp.*, and *Candida pseudotropicalis*. Furthermore, analysis of the presence of known bacteriocin encoding genes showed that the genes for various enterocins were present. Although in some strains more than one enterocin gene was detected, there was no correlation between the number of enterocin genes and the antimicrobial spectrum. Nevertheless, in order to characterize the strains that could be safely used as starter cultures in functional food, the frequency of virulence determinants and antibiotic resistance, as well as the synthesis of biogenic amines, was analyzed. The results show that the presence of virulence determinants and antibiotic resistance is strain dependent and region specific. In addition, a large percentage of the strains have the ability to decarboxylate tyrosine and other amino acids. Such capacity for decarboxylation of amino acids limits the use of the strains in the food industry. Based on these results, it can be concluded that enterococci isolated from animal food must be viewed with particular caution because they are reservoirs of genes for antibiotic resistance and virulence.

Chapter 8 – *Enterococcus faecalis* is a commensal bacterium inhabiting the gastrointestinal tract of humans. Interestingly, although it is not clear whether *E. faecalis* is part of the oral cavity microbiome, it is frequently recovered from root canal infections. Specifically, it is the major pathogen found in persistent infections associated with root canal treatment failure. Moreover, *E. faecalis* is one of the leading multidrug resistant nosocomial pathogens, causing infective endocarditis, and participating in urinary tract, wound, and device-device-related infections. The present chapter discusses *E. faecalis* virulence factors contributing to its high prevalence in nosocomial infections and root canal post treatment disease, including its ability to compete with other microorganisms, its cell to cell communication, its ability to invade various tissues, resist nutritional deprivation, facilitate the adherence of host cells and extracellular matrix, produce an immunomodulatory effect and cause toxin-mediated damage. Antiseptic techniques, conventional as well as novel, to overcome *the* survival ability of *E. faecalis as well as* virulence factors, are discussed in detail.

In: *Enterococcus faecalis*
Editor: Henry L. Mack

ISBN: 978-1-63321-049-3
© 2014 Nova Science Publishers, Inc.

Chapter 1

VIRULENCE FACTORS OF *ENTEROCOCCUS FAECALIS*: THE PROMOTERS FOR THEIR PATHOGENICITY

Joana Barbosa, Sandra Borges and Paula Teixeira[*]
CBQF – Centro de Biotecnologia e Química Fina, Escola Superior de Biotecnologia,
Centro Regional do Porto da Universidade Católica Portuguesa,
Porto, Portugal

ABSTRACT

Enterococci are natural inhabitants of the intestinal microbiota of humans and animals. They can survive in many environments contaminated with human or animal faecal materials (such as water, soils receiving and any surfaces) as well as food products derived from animals. They are important causes of community-acquired and nosocomial infections, such as urinary tract infections, endocarditis and bacteraemia. Among enterococcal diseases, the majority are associated with *Enterococcus faecalis* species. Several factors are associated with a high risk of acquiring enterococcal infections, including antimicrobial resistance and expression of virulence determinants. The most known virulence factors are enterococcal surface protein, aggregation substance, cell wall adhesins, sex pheromones, production of extracellular superoxide, production of hydrolytic enzymes (such as gelatinase), secretion of cytolysin, production of biofilm, among others.

To cause infection, *E. faecalis* must have virulence traits which allow the infecting strains to colonize and invade host tissue and translocate through epithelial cells and evade the host's immune response.

This review explores the importance of the presence of virulence factors for understanding the potential pathogenic activity of *E. faecalis*.

---

[*] Corresponding author: CBQF – Centro de Biotecnologia e Química Fina, Escola Superior de Biotecnologia, Centro Regional do Porto da Universidade Católica Portuguesa, Rua Dr. António Bernardino Almeida, 4200-072 Porto, Portugal; Tel +351 225 580 001; Fax: +351 225 580 111; e-mail: pcteixeira@porto.ucp.pt.

1. INTRODUCTION

Enterococci are natural inhabitants of the intestinal microbiota of humans and animals (Murray, 1990). Due to their ability to grow and survive under severe environmental conditions (Giraffa, 2002; Kayser, 2003; Varela et al., 2013), they can be present in many environments contaminated with human or animal faecal materials (such as water, soils receiving and any surfaces) as well as food products derived from animals (Barbosa et al., 2010; Pesavento et al., 2014). A small number of enterococci are also present in oropharyngeal secretions, vaginal secretions and on the skin, especially in the perineal area and, consequently, they can be considered as normal inhabitants of the human organism (Kayser, 2003).

Enterococcus faecalis is frequently the predominant species in the gastrointestinal tract, but in some cases the numbers of *E. faecium* are higher than those of *E. faecalis* (Hufnagel et al., 2007; Klein, 2003). Numbers of *E. faecalis* in human feces range from 10^5 to 10^8 CFU/g in contrast with 10^4 to 10^5 CFU/g for *E. faecium* (Jett et al. 1994; Leclerc et al. 1996).

Enterococci are responsible for several infections, either community-acquired or nosocomial infections, such as urinary tract infections, endocarditis, hepatobiliary sepsis, surgical wound infections, neonatal sepsis and bacteraemia (Poh et al., 2006). Enterococcal infections mainly occur in elderly and immuno-compromised individuals (Vu and Carvalho, 2011). Among these, approximately 80-90% are associated with *E. faecalis* (Huycke et al., 1998; Moellering, 1992). The majority of infections are nosocomial (Giraffa, 2002; Gómez-Gil et al., 2009), being the urinary tract infections the most common, which are originated by the use of urinary catheters from patients. Succeeding, polymicrobial infection (that appear from intraabdominal and pelvic infections) and bacteriemia are the second and third most common infections, respectively. Endocarditis can occurs afterwards other enterococcal infections and are associated with a high morbidity and mortality. They are also associated with dental infections, since they may colonize the oral cavity (reviewed in Vu and Carvalho, 2011).

The ability to cause infections in humans and the growth capacity of these bacteria in most habitats has resulted in their emergence as nosocomial pathogens worldwide (Franz et al., 1999; Miranda et al., 2001; Hummel et al., 2007) and determines an urgent need to know more about the significance of enterococci.

2. VIRULENCE FACTORS

Pathogens produce virulence factors in order to help their colonization, immunoevasion and immunosuppression of hosts. The presence of virulence factors could be associated with human disease (Jett et al., 1994; Johnson, 1994; Stiles and Holzapfel, 1997; Franz et al., 1999; Manero and Blanch, 1999; Giraffa, 2002).

Table 1 summarizes some of the virulence factors produced by *E. faecalis* and that will be discussed throughout this review.

2.1. Adherence to Host Tissues

For gastrointestinal inhabitants, such as enterococci, adherence is a crucial step in the infection process (Jett et al., 1994).

2.1.1. Enterococcal Surface Protein

The enterococcal surface protein (Esp) is a high molecular weight extracellular surface protein, which contributes to colonization and persistence of enterococci. This protein has been found in several species of enterococci, despite it was originally identified in *E. faecalis* (Shankar et al., 1999). This protein could be responsible for adherence of *Enterococcus* spp. to urinary bladder which could promote the urinary tract infections (Shankar et al., 2001).

The *esp* gene is unusually large consisting of 5622 nucleotides capable of encoding a primary translation product of 1873 amino acids, with a theoretical molecular mass of approximately 202 KDa (Shankar et al., 1999; Waar et al., 2002). This gene is more frequently found in clinical isolates of *E. faecalis* than non-clinical isolates, revealing the importance of Esp in nosocomial infections (Shankar et al., 1999).

Table 1. Virulence factors of *E. faecalis* and their function

Factor	Gene	Function	Reference
Enterococcal surface protein	*esp*	Adhesion and colonization	Shankar et al., 1999
Aggregation substance	*asp1*, *asc10* and *asa1*	Adhesion, colonization and resistance to host defense	Dunny et al., 1978; Kreft et al., 1992; Shankar et al., 2002
Collagen-adhesion protein	*ace*	Adhesion	Nallapareddy et al. 2000
Cell wall adhesins EfaA	*efaA*	Colonization	Lowe et al., 1995
Capsular polysaccharide	*epa* and *cps*	Resistance to host defense	Hancock and Gilmore, 2002; Xu et al., 1998
Sex pheromones	pheromone-inducible plasmids	Induction of inflammation	Dunny et al., 1978; Clewell and Weaver, 1989
Gelatinase and Serine protease	*gelE* *sprE*	Tissue damage	Mäkinen et al., 1989; Qin et al., 2000 Su et al., 1991
Cytolysin	*cylA-M*	Tissue damage and Inhibition of other bacteria	Gilmore et al., 1994; Haas and Gilmore, 1999

Esp also contributes to biofilm formation of *E. faecalis* on abiotic surfaces (Toledo-Arana et al., 2001). Studies have shown that rupture of the *esp* gene affects the capacity of *E. faecalis* to form biofilms. *Enterococcus faecalis* strains which do not possess this gene were able to form biofilms after receiving this gene by plasmid transfer (Latasa et al., 2006).

This protein possess an N-terminal signal sequence and a variable N-terminal domain of around 700 amino acids, followed by the repeated domains A, B and C, whose number differs among isolates (Leavis et al., 2004; Shankar et al., 1999). It also possesses the C-terminal end formed by a [Y/F]PXTG motif, which allow the protein to remain on the cell wall. It was suggested that N-terminal domain confer the ability of biofilm formation of Esp (Tendolkar et al., 2005).

The Esp exhibits global structural similarity to *Streptococcus pyogenes* R28 (Stalhammar-Carlemalm et al., 1999) and to the *Staphylococcus aureus* biofilm-associated protein BAP (Cucarella et al., 2001).

2.1.2. Aggregation Substance

Aggregation substance is a pheromone-inducible surface protein, found in *E. faecalis* strains (Eaton and Gasson, 2001), which facilitates plasmid transfer (Clewell, 1993). This protein is essential for high-efficiency conjugation of sex pheromone plasmids and also acts as a virulence factor during host infection, having a number of different functions like adherence to host tissues (Sartingen et al., 2000), adhesion to fibrin and increased cell surface hydrophobicity (Hirt et al., 2000) and resistance to killing by polymorphonuclear leukocytes and macrophages (Rakita et al., 1999). This protein improves the adherence to renal cells and heart endocardial cells by bacteria (Kreft et al., 1992). The ability conferred by this protein to adhere to and invade cells derived from the colon and duodenum, underline the possibility of the passage of *E. faecalis* across the cell wall, leading a systemic infection (Sartingen et al., 2000).

The aggregation protein is encoded by the genes *asp1*, *asc10* and *asa1* from pheromone-inducible conjugative plasmids pPD1, pCF10 and pAD1, respectively, and other enterococci secrete an oligopeptide autocrine pheromone that induces these genes expression (Chandler et al., 2005; Kozlowicz et al., 2006; Shankar et al., 2002). Different domains organize this protein: a lipoteichoic acid aggregation domain, a variable region, a central aggregation domain and two Arg-Gly-Asp motifs. The presence of aggregation protein at the surface allows the contact between the cells which facilitate the transfer of virulence plasmids (Hendrickx et al., 2009).

2.1.3. Collagen-Adhesion Protein

Ace is a collagen-adhesion protein from *E. faecalis* and it was the first MSCRAMM (Microbial Surface Component Recognizing Adhesive Matrix Molecule adhesin of collagen from enterococci) described among enterococci and mediates binding to collagen (types I and type IV), laminin and dentin (Rich et al., 1999).

The protein Ace is encoded by *ace* gene and the diversity of this gene from different strains of *E. faecalis* was firstly studied by Nallapareddy et al. (2000).

During infection, Ace can be expressed in commensal as well as pathogenic *E. faecalis* isolates. It is thought to play a role in the pathogenesis of endocarditis, since it was detected in 90% of enterococcal endocarditis patient sera samples (Koch et al., 2004).

2.1.4. Cell Wall Adhesins EfaA

The cell wall adhesins efaAfm and efaAfs of *E. faecium* and *E. faecalis*, respectively, are considered as potential virulence determinants. The antigen A gene, encoding a 37-kDa dominant antigen, was previously identified by Lowe et al. (1995) using serum from a patient with *E. faecalis* endocarditis. The associated protein EfaA revealed 55 to 60% homology in amino acid sequence to a group of streptococcal adhesins (Kayaoglu and Ørstavik, 2004).

The *efaA* gene is the third gene of the *efaCBA* operon, being EfaA the lipoprotein component of the ABC-type transporter. Low et al. (2003) described that expression of the endocarditis-associated virulence factor EfaA is manganese-regulated and the *efaCBA* operon of *E. faecalis* depending on EfaR – a Mn^{2+} responsive transcriptional regulator. A recent study proved that EfaR is a relevant modulator of *E. faecalis* virulence and binding Mn^{2+} homeostasis to enterococcal pathogenicity (Abrantes et al., 2013).

2.2. Capsular Polysaccharide

The capsule is an important element in the pathogenicity of *E. faecalis*, since it confers resistance to phagocytosis, allowing the pathogen to escape to host immune system (Hancock and Gilmore, 2002). Serotyping of *E. faecalis* is based in the differences between capsular polysaccharide antigens as well as surface antigens (Hufnagel et al., 2004).

The synthesis of capsular polysaccharide is performed by two loci: *epa* (enterococcal polysaccharide antigen) and *cps* operons. The *cps* operon comprises 11 open reading frames (*cpsA* to *cpsK*) and *epa* operon encode a rhamnose-containg polysaccharide (Hancock and Gilmore, 2002; Xu et al., 1998).

The specific mechanism of how the presence of capsule increases the pathogenicity of *E. faecalis* is still unknown.

2.3. Sex Pheromones

The first description of the sex pheromone system of *E. faecalis* was provided by the collaborators of D.B. Clewell, during studies on the conjugative transfer of a special class of plasmids (Dunny et al., 1978). This conjugative plasmid system allows the efficient transfer of genes, such as antibiotic resistance genes and other genes that increase the capability to promote disease (Clewell et al., 1974; Clewell and Dunny, 2002).

The plasmid transfer is initiated as a response to an extremely specific sex pheromone peptide secreted by a potential receptor. First, a receptor cell produces a sex pheromone corresponding to the plasmid that it does not have. Then, the donor cell produces the aggregation substance, allowing the conjugative transfer of the replicated plasmid. As soon as the plasmid is acquired by the receptor cell, the production of the sex pheromone is interrupted. Moreover, the receptor cell continues producing other sex pheromones corresponding to other plasmids (Clewell and Weaver, 1989).

Sex pheromones are small hydrophobic peptides (7 or 8 amino acids), chromosomally encoded, with signal like function (Clewell and Weaver, 1989). The pCF10, pAD1 and pPD1 are most well studied pheromone-inducible plasmids of *E. faecalis* and have similar regulatory mechanisms (Clewell et al., 2000). These plasmids can also carry antibiotic resistances, for example, the plasmid pCF10 encodes aggregation substance and tetracycline resistance, being important for both virulence and conjugation (Chen et al., 2008).

2.4. Production of Extracellular Superoxide

Superoxide anion is a reactive oxygen radical responsible for cell and tissue injuries, since it destroys biological compounds such as lipids, proteins and nucleic acids (Cross et al., 1987).

It has been demonstrated that *E. faecalis* is the only intestinal commensal producing extracellular superoxide, proposing an association of virulence and extracellular superoxide production (Huycke et al., 2002; Huycke et al., 1996; Wang and Huycke, 2007).

To our knowledge the exact function of superoxide in the host tissue is still unknown.

2.5. Production of Hydrolytic Enzymes

2.5.1. Gelatinase and Serine Protease

Gelatinase (Gel) is an extracellular metallo-endopeptidase responsible for the hydrolysis of small biologically active peptides such as gelatin, casein, collagen and other bioactive peptides. This protein has a molecular weight of 31.5 kDa, an isoelectric point of 4.6, strongly hydrophobic, and pH optimum of 6.8 (Mäkinen et al., 1989). Serine protease (SprE) has a molecular weight of 26 kDa and shares homology with *S. aureus* V8 protease (Qin et al., 2000). These proteases seem to have an important contribution to enterococcal pathogenesis because degrading the host tissues they supply nutrient sources.

The gene encoding gelatinase is named *gelE* and appears to be located in the chromosome (Su et al., 1991). The gene *sprE* encode serine protease and is located in an operon together with the *gelE* gene. Transcription of *gelE* and *sprE* is regulated by the quorum-sensing system encoded by the *fsr* (faecal streptococci regulator) locus, containing four genes: *fsrA, fsrB, fsrD* and *fsrC* (Lopes et al., 2006; Nakayama et al., 2006; Qin et al., 2000).

The interaction of GelE and SprE is responsible for the regulation of autolysis of *E. faecalis* cells, causing the release of extracellular DNA, which can be one of the causes to enhance biofilm production (Thomas et al., 2008).

Gelatinase is most frequently found in clinical isolates than in isolates collected from faeces of health individuals (Coque et al., 1995). Several studies using animal models have been shown the importance of gelatinase as the principal mediator of pathogenesis of *E. faecalis*, such as endophthalmitis (Suzuki et al., 2008), endocarditis (Thurlow et al., 2010) and in chronic intestinal inflammation (Steck et al., 2011).

2.5.2. Hyaluronidase

Bacterial hyaluronidase is an enzyme capable of breaking down the substrate hyaluronate (linear unsulfated glycosaminoglycan polymer founded in many body tissues and fluids of higher organisms such as umbilical cord, synovial fluid, cartilage, brain and muscle) and is produced by a number of pathogenic Gram-positive bacteria that initiate infections at the skin or mucosal surfaces (Hynes and Walton, 2000). This enzyme is able to depolymerize the mucopolysaccharide moiety of connective tissues in order to facilitate the invasion of enterococci even as the propagation of their toxins (Fisher and Philips, 2009). Hyaluronidase can also provide nutrients for enterococci, because they can transport and metabolize the disaccharides derived from the degradation of substrate hyaluronate (Kayaoglu and Orstavik, 2004).

Hyaluronidase is expressed by the gene *hyl*, which is located into the chromosome (Fisher and Philips, 2009).

The production of hyaluronidase by *E. faecalis* has been related with the injury of dentinal and periapical tissues (Halkai et al., 2012).

2.6. Secretion of Cytolysin

Cytolysin is a post-translationally modified toxic protein, only found in *E. faecalis*, which causes a β-hemolytic reaction on certain blood erythrocytes and, in addition, has a bactericidal effect (Haas and Gilmore, 1999). Besides erythrocytes, cytolysin, present in *E. faecalis* strains, can also lyse polymorphonuclear neutrophils and macrophages (Miyazaki et al., 1993). This compound forms pores in the cytoplasmic membrane of target bacterial cells (Gilmore et al., 1994).

The genes encoding for cytolysin are located into an operon that can either be transported on pheromone-responsive plasmids or integrated into the chromossome (Koch et al., 2004). Gilmore et al. (1994) showed that production of cytolytic activity by *E. faecalis* requires expression of eight reading frames: *cylM* (gene involved in modification), *cylB* (secretion), *cylA* (activation), *cylL$_L$* and *cylL$_s$*. The lytic factor precursors CylL$_L$ (the long subunit) and CylL$_S$ (the short subunit) are ribosomally synthesized and modified post-translationally by CylM. The modified peptides are then cleaved by proteolitic enzymes and secreted from the cell by CylB, an ATP-binding cassette transporter. The secreted peptide subunits CylL$_L$' and CylL$_S$' are further cleaved and activated extracellularly by CylA, a serine protease. "CylL$_L$" and CylL$_S$" are both required for the lysis of human or target bacterial cells. The genes *cylR1* and *cylR2* encode two regulator proteins of the operon. The *cylI* gene encodes a membrane protein responsible for immunity system function (Clewell, 2007; Haas and Gilmore, 1999).

2.7. Biofilm Formation by Enterococci

A biofilm is described as a complex aggregation of microorganisms attached on various biotic and abiotic surfaces enveloped in a matrix composed by exopolysacharides, proteins and nucleic acids (Hall-Stoodley et al., 2004; Jefferson, 2004). In biofilm production, several stages are needed. Firstly, an adhesion to the surface is required (**adhesion phase**), occurring **microcolony formation (young biofilm production)**. Then, extracellular polymeric compounds are secreted and finally the development to a mature biofilm and dissociation (Costerton et al., 1995). Whereas bacteria may adhere to a surface within minutes, it is generally assumed that true biofilms take hours or days to develop (Hood and Zottola, 1995). Enterococci have been associated with the production of biofilms in root canals, on different contact materials (Mohamed et al., 2004) and on various indwelling medical devices, such as urethral stents (Keane et al., 1994) and intravascular catheters (Sandoe et al., 2003).

Biofilm production is regulated by quorum sensing, which is the regulation of bacterial gene expression in response to changes in cell-population density. Quorum sensing bacteria produce extracellular signal molecules called autoinducers (Miller and Bassler, 2001).

Several genes have been identified in *E. faecalis* as being involved in biofilm formation, such as the *esp*, *gelE*, *ebp*, *bee*, *epa*, among others.

Enterococcal surface protein has been associated with biofilm formation ability among clinical *E. faecalis* isolates. Toledo-Arana (2001) demonstrated that 87 out of 93 *E. faecalis* strains that possessed *esp* gene were able to form biofilm. So it seems that although be an important factor, Esp is not essential to biofilm formation.

Proteases also play an important role in biofilm ability. Hancock and Perego (2004) showed that a mutagenesis in the *fsr* operon results in a decrease on biofilm formation and gelatinase production. The deletion of *gelE* gene resulted in deficiency of autolysis and biofilm (Thomas et al., 2008).

Another essential factor for biofilm production is the formation of pili, which is associated with *ebp* (endocarditis- and biofilm-associated pili) operon. This operon is constituted by three genes: *ebpA*, *ebpB* and *ebpC* and this locus is positively regulated by the downstream *ebpR* gene, which is regulated by *fsr* (Bourgogne et al., 2007). Tendolkar et al. (2006) identified a cluster of biofilm-associated genes, designated *bee* (biofilm enhancer in *Enterococcus*) locus, which contain five genes encoding three cell wall-anchored proteins and two possible sortases, involved in Bee pilus formation. The *bee* locus is carried on a conjugative plasmid and other *E. faecalis* strains could receive it, resulting in biofilm production improvement by the transconjugants (Tendolkar et al., 2006).

The *epa* gene is involved in cell-wall associated polysaccharides biosynthesis of *E. faecalis*. Mohamed et al. (2004) studied an *E. faecalis epa* mutant and observed a reduction on biofilm production of approximately 73%, indicating that *epa* gene is involved in biofilm production.

The biofilm production is a strategy to survive harsh conditions providing protection for bacteria and becoming difficult to eradicate. For instance, in biofilm state bacteria are more resistant to phagocytosis, as well as higher concentrations of antibiotics are required to eliminate them (Lewis et al., 2001). This antibiotic tolerance in biofilm state and intrinsic antibiotic resistance of *E. faecalis*, could led to the dissemination of antibiotic resistance genes (Elhadidy and Elsayyad, 2013).

3. ANTIBIOTIC RESISTANCE

Enterococci are resistant to a wide variety of antibiotics. This feature allows enterococci to survive in the hospital environment where antibiotics are used and the hospital provides the opportunity for dissemination of resistant organisms (Murray, 1990; Leclercq, 1997). This resistance can be both intrinsic, mediated by genes located on the chromosome – a characteristic present in all or most of the strains of enterococci species (Moellering and Krogstad, 1979), - and acquired, mediated by genes residing on plasmids or transposons (Murray, 1990).

Enterococci are intrinsic resistant to cephalosporins, sulphonamides, lincosamides, many β-lactams and to low levels of aminoglycosides (Moellering, 1990; Murray, 1990; Leclercq, 1997; Morrison et al., 1997). The acquired genetic determinants confer resistance to all classes of antimicrobials, including chloramphenicol, tetracyclines, erythromycin, rifampicin, ampicillin and glycopeptides. The main concern is that the genes coding for all these antibiotic resistance traits could be transferred by pheromone mediated, conjugative plasmids or transposons to other enterococci or even to more virulent pathogens (Morrison et al., 1997; Schwarz et al., 2001).

The treatment of enterococcal infections in human medicine can be very difficult when intrinsic resistance exists, but it can be successfully solved by combinations of cell-wall-active antibiotics with aminoglycosides, like gentamicin or streptomycin (Moellering, 1990; Murray, 1990). The glycopeptide antibiotics (vancomycin and teicoplanin) are important reserve antibiotics in case of resistance or allergy to penicillins and high-level resistance to aminoglycosides (Peters et al., 2003). Nitrofurantoin is effective for the treatment of enterococcal urinary tract infections, including many caused by vancomycin resistant enterococci (VRE) strains (Zhanel et al., 2001). Chloramphenicol (Norris et al., 1995) and tetracycline (Howe et al., 1997), in various combinations, have also been used to treat VRE infections.

The resistance to glycopeptides is another clinically relevant characteristic (Endtz et al., 1999). Almost all Gram-positive bacteria are susceptible to their activity. On the contrary, glycopeptides are relatively large water-soluble molecules that cannot penetrate the lipid outer membrane of Gram-negatives (Houben, 2003). Six phenotypes have been reported for enterococci: vanA, vanB, vanC (Moellering, 1991; Moellering and Gold, 1996), vanD (Arthur et al., 1993), vanE (Fines et al., 1999) and vanG (Messi et al., 2006). The most frequent type isolated from clinical cases is vanA, which is characterized by the acquired inducible high level-resistance to vancomycin (minimal inhibitory concentration (MIC) > 256 µg/ml) and teicoplanin (MIC > 16 µg/ml). This is the type most commonly found in *E. faecium* and *E. faecalis* (Arthur and Courvalin, 1988; Arthur et al., 1993). The *vanB* genes confer lower level acquired inducible resistance to vancomycin (MIC>64 µg/ml), but not to teicoplanin and are only found in *E. faecium* and *E. faecalis* (Quintiliani *et al.*, 1993; Arthur et al., 1996). *Enterococcus gallinarum* (*vanC*1), *E. casseliflavus* (*vanC*2) and *E. flavescens* (*vanC*3) have intrinsic low-level resistance to both vancomycin and teicoplanin (MIC 4-16 µg/ml) and the vanC type appears to be an intrinsic property of the motile species (Dukta-Malen et al., 1995; Leclercq and Courvalin, 1997). The vanD type is characterized by the acquired intermediate level resistance to both glycopeptides (Arthur et al., 1993; Perichon et al., 1997), vanE is responsible for low-level acquired resistance in *E. faecalis* (Fines et al., 1999) and *vanG* genes confer low-level resistance to vancomycin only (McKessar et al., 2000; Depardieu et al., 2003).

This high antibiotic resistance is a cause of concern, however, the antibiotic resistance alone cannot explain the pathogenicity of enterococci in the absence of other virulence factors (Giraffa, 2002).

CONCLUSION

Enterococci are microorganisms capable to survive to several adverse conditions, which means that they can be disseminated different sources (environment, food and clinical) and cause disease. In order to become pathogenic, they must be able to adhere, colonize, and resist/evade the immune host system. Enterococci produce several virulence factors which allow them to accomplish all the infection steps, detailed throughout this review.

The presence of these virulence factors in *E. faecalis* strains, together with their intrinsic and acquired antibiotic resistance, were the essential requisites to turn this intestinal commensal bacteria into a virulent pathogen, causing so many nosocomial infections.

REFERENCES

Abrantes, M. C., Kok, J., & Lopes, M. de F. (2013). EfaR Is a Major Regulator of *Enterococcus faecalis* Manganese Transporters and Influences Processes Involved in Host Colonization and Infection. *Infection and Immunity, 81*, 935-944.

Arthur, M., & Courvalin, P. (1988). Genetics and mechanisms of glycopeptide resistance in enterococci, *Antimicrobial Agents and Chemotherapy, 319*, 157-161.

Arthur, M., Molinas, C., Depardieu, F., & Courvalin, P. (1993). Characterization of Tn1546, a tn3-related transposon conferring glycopeptide resistance by synthesis of depsipeptide peptidoglycan precursors in *Enterococcus faecium* BM4147. *Journal of Bacteriology, 175*, 117-127.

Arthur, M., Reynolds, P.E., & Courvalin, P. (1996). Glycopeptide resistance in enterococci. *Trends in Microbiology, 4*, 401-407.

Barbosa, J., Gibbs, P. A., & Teixeira., P. (2010). Virulence factors among enterococci isolated from traditional fermented meat products produced in the North of Portugal. *Food Control, 21*, 651-656.

Bourgogne, A., Singh, K.V., Fox, K.A., Pflughoeft, K.A., Murray, B.E., & Garsin, D.A. (2007). EbpR is important for biofilm formation by activating expression of the endocarditis and biofilm-associated pilus operon (ebpABC) of *Enterococcus faecalis* OG1RF. *Journal of Bacteriology, 189*, 6490–6493.

Chandler, J.R., Hirt, H., & Dunny, G.M. (2005). A paracrine peptide sex pheromone also acts as an autocrine signal to induce plasmid transfer and virulence factor expression in vivo. *Proceedings of the National Academy of Sciences of the United States of America, 102*, 15617–15622.

Chen, Y., Zhang, X., Manias, D., Yeo, H.-J., Dunny, G.M., & Christie, P.J. (2008). *Enterococcus faecalis* PcfC, a spatially localized substrate receptor for type IV secretion of the pCF10 transfer intermediate. *Journal of Bacteriology, 190*, 3632–3645.

Clewell, D.B. (1993). Bacterial sex pheromone-induced plasmid transfer. *Cell, 73*, 9-12.

Clewell, D. B. (2007). Properties of *Enterococcus faecalis* plasmid pAD1, a member of a widely disseminated family of pheromone-responding, conjugative, virulence elements encoding cytolysin. *Plasmid, 58*, 205–227.

Clewell, D. B., An, F. Y., Flannagan, S. E., Antiporta, M. & Dunny, G. M. (2000). Enterococcal sex pheromone precursors are part of signal sequences for surface lipoproteins. *Molecular Microbiology, 35*, 246–247.

Clewell, D.B., & Dunny, G.M. (2002). Conjugation and genetic exchange in enterococci. In: Gilmore, M.S., Clewell, D.B., Courvalin, P., Dunny, G.M., Murray, B.E., Rice, L.B. (Eds.). *The Enterococci: Pathogenesis, Molecular Biology, and Antibiotic Resistance.* ASM Press, Washington, DC, 265–300.

Clewell, D.B., & Weaver, K.E. (1989). Sex pheromones and plasmid transfer in *Enterococcus faecalis. Plasmid, 21*, 175-184.

Clewell, D.B., Yagi, Y., Dunny, G.M., & Schultz, S.K. (1974). Characterization of three plasmid deoxyribonucleic acid molecules in a strain of *Streptococcus faecalis*: identification of a plasmid determining erythromycin resistance. *Journal of Bacteriology, 117*, 283–289.

Coque, T.M., Patterson, J.E., Steckelberg, J.M., & Murray, B.E. (1995). Incidence of hemolysin, gelatinase, and aggregation substance among enterococci isolated from patients with endocarditis and other infections and from feces of hospitalized and community-based persons. *Journal of Infectious Diseases, 1715*, 1223-1229.

Costerton, J.W., Lewandowski, Z., Caldwell, D.E., Korber, D.R., & Lappin-Scott, H.M. (1995). Microbial biofilms. *Annual Review of Microbiology, 49*, 711–745.

Cross, C.E., Halliwell, B., Borish, E.T., Pryor, W.A., Ames, B.N., Saul, R.L., McCord, J.M., & Harman, D. (1987). Oxygen radicals and human disease. *Annals of Internal Medicine, 107*, 526-545.

Cucarella, C., Solano, C., Valle, J., Amorena, B., Lasa, I., & Penadés, J.R. (2001). Bap, a *Staphylococcus aureus* surface protein involved in biofilm formation. *Journal of Bacteriology, 9*, 2888–2896.

Depardieu, F., Bonora, M.G., Reynolds, P.E., & Courvalin, P. (2003). The *vanG* glycopeptide resistance operon from *Enterococcus faecalis* revisited. *Molecular Microbiology,* **50***, 931-948.

Dukta-Malen, S., Evers, S., & Courvalin, P. (1995). Detection of glycopeptide resistance genotypes and identification to the species level of clinically relevant enterococci by PCR. *Journal of Clinical Microbiology, 33*, 24-27. (Erratum, *33*:1434)

Dunny, G.M., Brown, B.L., & Clewell, D.B. (1978). Induced cell aggregation and mating in *Streptococcus faecalis*: evidence for a bacterial sex pheromone. *Proceedings of the National Academy of Sciences of the United States of America, 75*, 3479–3483.

Eaton, T.J., & Gasson, M.J. (2001). Molecular screening of *Enterococcus* virulence determinants and potential for genetic exchange between food and medical isolates. *FEMS Microbiology Letters, 67*, 1628-1635.

Elhadidy, M., & Elsayyad, A. (2013). Uncommitted role of enterococcal surface protein, Esp, and origin of isolates on biofilm production by *Enterococcus faecalis* isolated from bovine mastitis. *Journal of Microbiology, Immunology and Infection, 46*, 80-84.

Endtz, H.P., Van den Braak, N., Verbrugh, H.A., & Van Bekum, A. (1999). Vancomycin resistance *status quo* and *quo vadis. European Journal Clinical Microbiology & Infectious Diseases, 18*, 683-690.

Fisher, K., & Phillips, C. (2009). The ecology, epidemiology and virulence of *Enterococcus. Microbiology, 155*, 1749–1757.

Franz, C.M.A.P., Holzapfel, W.H., & Stiles, M.E. (1999). Enterococci at the crossroads of food safety? *International Journal of Food Microbiology, 47*, 1-24.

Gilmore, M.S., Segarra, R.A., Booth, M.C., Bogie, Ch.P., Hall, L.R., & Clewell, D.B. (1994). Genetic structure of *Enterococcus faecalis* plasmid pAD1-encoded cytolytic toxin system and its relationship to lantibiotic determinants. *Journal of Bacteriology, 176*, 7335-7344.

Giraffa, G. (2002). Enterococci from foods. *FEMS Microbiology Reviews, 26*, 163-171.

Gómez-Gil, R., Romero-Gómeza, M.P., García-Ariasa, A., Ubedab, M.G., Busseloc, M.S., Cisternac, R., Gutiérrez-Altésa, A., & Mingorance, J. (2009). Nosocomial outbreak of linezolid-resistant *Enterococcus faecalis* infection in a tertiary care hospital. *Diagnostic Microbiology and Infectious Disease, 65*, 175–179.

Haas, W., & Gilmore, M.S. (1999). Molecular nature of a novel bacterial toxin: the cytolysin of *Enterococcus faecalis*. *Medical Microbiology and Immunology, 187*, 183-190.

Halkai, R., Hegde, M.N., & Halkai, K. (2012). *Enterococcus faecalis* can survive extreme challenges – overview. *Nitte University Journal of Health Science, 2*, 49-53.

Hall-Stoodley, L., Costerton, J.W., & Stoodley, P. (2004). Bacterial biofilms: from the natural environment to infectious diseases. *Nature Reviews Microbiology, 2*, 95–108.

Hancock, L.E., & Gilmore M.S. (2002). The capsular polysaccharide of *Enterococcus faecalis* and its relationship to other polysaccharides in the cell wall. *Proceedings of the National Academy of Sciences of the United States of America, 99*, 1574–1579.

Hancock, L.E., & Perego, M. (2004). The *Enterococcus faecalis* fsr two-component system controls biofilm development through production of gelatinase. *Journal of Bacteriology, 186*, 5629–5639.

Hendrickx, A.P.A., Willems, R.J.L., Bonten, M.J.M., & van Schaik, W. (2009). LPxTG surface proteins of enterococci. *Trends in Microbiology, 17*, 423-430.

Hirt, H., Erlandsen, S.L., & Dunny, G.M. (2000). Heterologous inducible expression of *Enterococcus faecalis* pCF10 aggregation substance Asc10 in *Lactococcus lactis* and *Streptococcus gordonii* contributes to cell hydrophobicity and adhesion to fibrin. *Journal of Bacteriology, 182*, 2299-2306.

Hood, S.K., & Zottola, E.A. (1995). Biofilms in food processing. *Food Control, 6*, 9-18.

Houben, J.H. (2003). The potential of vancomycin-resistant enterococci to persist in fermented and pasteurized meat products. *International Journal of Food Microbiology, 88*, 11-18.

Howe, R. A., Robson, M., Oakhill, A., Cornish, J. M., & Millar, M. R. (1997). Successful use of tetracycline as therapy of an immunocompromised patient with septicaemia caused by a vancomycin-resistant *enterococcus*. *Journal of Antimicrobial Chemotherapy, 40*, 144–145.

Hufnagel, M., Hancock, L.E., Koch, S., Theilacker, C., Gilmore, M.S., & Huebner, J. (2004). Serological and genetic diversity of capsular polysaccharides in *Enterococcus faecalis*. *Microbiology, 42*, 2548–2557.

Hufnagel, M., Liese, C., Loescher, C., Kunze, M., Proempeler, H., Berner, R, & Krueger, M. (2007). Enterococcal colonization of infants in a neonatal intensive care unit: associated predictors, risk factors and seasonal patterns. *BMC Infectious Diseases, 7*, 107-118.

Hummel, A., Holzapfel, W.H., & Franz, C.M.A.P. (2007). Characterisation and transfer of antibiotic resistance genes from enterococci isolated from food. *Systematic and Applied Microbiology, 30*, 1-7.

Huycke, M.M., Abrams, V., & Moore, D.R. (2002). *Enterococcus faecalis* produces extracellular superoxide and hydrogen peroxide that damages colonic epithelial cell DNA. *Carcinogenesis, 23*, 529–536.

Huycke, M.M., Joyce, W., & Wack, M.F. (1996). Augmented production of extracellular superoxide production by blood isolates of *Enterococcus faecalis*. *The Journal of Infectious Diseases, 173*, 743 –746.

Huycke, M.M., Sahm, D.F., & Gilmore, M.S. (1998). Multiple-drug resistant enterococci: the nature of the problem and an agenda for the future. *Emerging Infectious Diseases Journal, 4*, 239-249.

Hynes, W.L., & Walton, S.L. (2000). Hyaluronidases of Gram-positive bacteria. *FEMS Microbiology Letters, 183*, 201-207.

Jefferson, K.K. (2004). What drives bacteria to produce a biofilm? *FEMS Microbiology Letters, 236*, 163–173.

Jett, B., Huycke, M.M., & Gilmore, M.S. (1994). Virulence of enterococci. *Clinical Microbiology Reviews, 7*, 462-478.

Jonhson, A.P. (1994). The pathogenicity of enterococci. *Journal of Antimicrobial Chemotherapy, 33*, 1083-1089.

Kayaoglu, G., & Ørstavik, D. (2009). Virulence factors of *Enterococcus faecalis*: relationship to endodontic disease. *Criticial Reviews in Oral Biology & Medicine, 15*, 308-320.

Kayser, F.H. (2003). Safety aspects of enterococci from the medical point of view. *International Journal of Food Microbiology, 88*, 255-262.

Klein, G. (2003). Taxonomy, ecology and antibiotic resistance of enterococci from food and the gastro-intestinal tract. *International Journal of Food Microbiology, 88*, 123–131.

Koch, S., Hufnagel, M., Theilacker, C., & Huebner, J. (2004). Enterococcal infections: host response, therapeutic, and prophylactic possibilities. *Vaccine, 22*, 822–830.

Kozlowicz, B.K., Dworkin, M., & Dunny, G.M. (2006). Pheromone-inducible conjugation in *Enterococcus faecalis*: a model for the evolution of biological complexity? *International Journal of Medical Microbiology, 296*, 141–147.

Kreft, B., Marre, R., Schramm, U., & Wirth, R. (1992). Aggregation substance of *Enterococcus faecalis* mediates adhesion to cultured renal tubular cells. *Infection and Immunity, 60*, 25-30.

Latasa, C., Solano, C., Penadés, J. R., & Lasa, I. (2006). Biofilm associated proteins. *Comptes Rendus Biologies, 329*, 849-857.

Leclerc, H., Devriese, L.A., & Mossel, D.A.A. (1996). Taxonomical changes in intestinal (faecal) enterococci and streptococci: Consequences on their use as indicators of faecal contamination in drinking water. *Journal of Applied Microbiology, 81*, 459-466.

Leclercq, R. (1997). Enterococci acquire new kinds of resistance. *Clinical Infectious Diseases, 24* (Suppl. 1): S80-S84.

Leclercq, R., & Courvalin, P. (1997). Resistance to glycopeptides in enterococci. *Clinical Infectious Diseases, 24*, 545-554.

Leavis, H., Top, J., Shankar, N., Borgen, K., Bonten, M., van Embden, J., & Willems, R.J.L. (2004). A novel putative enterococcal pathogenicity island linked to the *esp* virulence gene of *Enterococcus faecium* and associated with epidemicity.*Journal of Bacteriology, 186*, 672–682.

Lewis, K. (2001). Riddle of biofilm resistance. *Antimicrobial Agents and Chemotherapy, 45*, 999–1007.

Lopes, M.F.S., Ribeiro, T., Abrantes, M., Marques, J.J.F., Tenreiro, R., & Crespo, M.T.B. (2005). Antimicrobial resistance profiles of dairy and clinical isolates and type strains of enterococci. *International Journal of Food Microbiology, 103*, 191-198.

Low, Y.L., Jakubovics, N.S., Flatman, J.C., Jenkinson, H.F., & Smith, A.W. (2003). Manganese-dependent regulation of the endocarditis-associated virulence factor EfaA of *Enterococcus faecalis*. *Journal of Medical Microbiology, 52*, 113–119.

Lowe, A.M., Lambert, P.A., & Smith, A.W. (1995). Cloning of an *Enterococcus faecalis* endocarditis antigen: homology with adhesins from some oral streptococci. *Infection and Immunity, 63*, 703 706.

Mäkinen, P.L., Clewell, D.B., An, F., & Mäkinen, K.K. (1989). Purification and substrate specificity of a strongly hydrophobic extracellular metalloendopeptidase ("gelatinase") from *Streptococcus faecalis* (strain OG1-10). *Journal of Biological Chemistry, 264*, 3325–3334.

Manero, A., & Blanch, A.R. (1999). Identification of *Enterococcus* spp. with a Biochemical Key. *Applied and Environmental Microbiology, 65*, 4425-4430.

McKessar, S.J., Berry, A.M., Bell, J.M., Turnidge, J.D., & Paton, J.C. (2000). Genetic characterization of *vanG*, a novel vancomycin resistance locus of *Enterococcus faecalis*. *Antimicrobial Agents and Chemotherapy, 44*, 3224-3228.

Messi, P., Guerrieri, E., Niederhäusern, S., Sabia, C., & Bondi, M. (2006). Vancomycin-resistant enterococci (VRE) in meat and environmental samples. *International Journal of Food Microbiology, 107*, 218-222.

Miller, M.B., & Bassler, B.L. (2001). Quorum sensing in bacteria. *Annual Review of Microbiology, 55*, 165–199.

Miranda, G., Lee, L., Kelly, C., Solórzano, F., Leaños, B., Muñoz, O., & Patterson, J.E. (2001). Antimicrobial Resistance from Enterococci in a Pediatric Hospital. Plasmids in *Enterococcus faecalis* Isolates with High-Level Gentamicin and Streptomycin Resistance. *Archives of Medical Research, 32*, 159-163.

Moellering, R.C. (1990). The enterococci: an enigma and a continuing therapeutic challenge. *European Journal of Clinical Microbiology & Infectious Diseases, 9*, 73-74.

Moellering, R.C. (1991). The garrod lecture. The *Enterococcus*: a classic example of the impact of antimicrobial resistance on therapeutic options. *Journal of Antimicrobial Chemotherapy, 28*, 1-12.

Moellering, R.C. (1992). Emergence of *Enterococcus* as a significant pathogen. *Clinical Infectious Diseases, 14*, 1173-1178.

Moellering, R.C., & Gold, H.S. (1996). Antimicrobial-drug resistance. *The New England Journal of Medicine, 335*, 1445-1453.

Moellering, R.C., & Krogstad, D.J. (1979). In: Schlessinger, D. (Ed.), Antibiotic Resistance in Enterococci. *American Society for Microbiology*, Washington DC, pp. 293–298.

Mohamed, J.A., Huang, W., Nallapareddy, S.R., Teng, F., & Murray, B.E. (2004). Influence of origin of isolates, especially endocarditis isolates, and various genes on biofilm formation by *Enterococcus faecalis*. *Infection and Immunity, 72*, 3658–3663.

Morrison, D., Woodford, N., & Cookson, B. (1997). Enterococci as emerging pathogens of humans. *Journal of Applied Microbiology Symposium Supplement, 83*, 89-99.

Murray, B. E. (1990). The life and Times of the *Enterococcus*. *Clinical Microbiological Reviews, 3*, 46-65.

Miyazaki, S., Ohno, A., Kobayashi, I., Uji, T., Yamaguchi, K., & Goto, S. (1993). Cytotoxic effect of hemolytic culture supernatant from *Enterococcus faecalis* on mouse polymorphonuclear neutrophils and macrophages. *Microbiology and Immunology, 37*, 265–270.

Nakayama, J., Chen, S., Oyama, N., Nishiguchi, K., Azab, E.A., Tanaka, E., Kariyama, R., & Sonomoto, K. (2006). Revised model for *Enterococcus faecalis* fsr quorum-sensing system: the small open reading frame fsrD encodes the gelatinase biosynthesis-activating pheromone propeptide corresponding to staphylococcal AgrD. *Journal of Bacteriology, 188*, 8321–8326.

Nallapareddy, S.R., Singh, K.V., Duh, R.-W., Weinstock, G.M., & Murray, B.E. (2000). Diversity of ace, a Gene Encoding a Microbial Surface Component Recognizing Adhesive Matrix Molecules, from Different Strains of *Enterococcus faecalis* and Evidence for Production of Ace during Human Infections. *Infection and Immunity, 68*, 5210-5217.

Norris, A.H., Reilly, J.P., Edelstein, P.H., Brennan, P.J., & Schuster, M.G. (1995). Chloramphenicol for the treatment of vancomycin-resistant enterococcal infections. *Clinical Infectious Diseases, 20,* 1137-1144.

Perichon, B., Reynolds, P., & Courvalin, P. 1997. *VanD*-type glycopeptide-resistant *Enterococcus faecium* BM4339. *Antimicrobial Agents and Chemotherapy, 41*, 2016-2018.

Pesavento, G., Calonico, C., Ducci, B., Magnanini, A., & Lo Nostro, A. (2014). Prevalence and antibiotic resistance of *Enterococcus* spp. isolated from retail cheese, ready-to-eat salads, ham, and raw meat. *Food Microbiology, 41*, 1-7.

Peters, J., Mac, K., Wichmann-Schauer, H., Klein, G., & Ellerbroek, L. (2003). Species distribution and antibiotic resistance patterns of enterococci isolated from food of animal origin in Germany. *International Journal of Food Microbiology, 88*, 311–314.

Poh, C.H., Oh, H.M.L., & Tan, A.L. (2006). Epidemiology and clinical outcome of enterococcal bacteraemia in an acute care hospital. *Journal of Infection, 52*, 383–386.

Qin, X., Singh, K.V., Weinstock, G.M., & Murray, B.E. (2000). Effects of *Enterococcus faecalis fsr* genes on production of gelatinase and a serine protease and virulence. *Infection and Immunity, 68***,** 2579-2586.

Quintiliani, R.Jr., Evers, S., & Courvalin, P. (1993). The *vanB* gene confers various levels of self-transferable resistance to vancomycin in enterococci. *The Journal of Infectious Diseases, 167*, 1220-1223.

Rakita, R.M., Vanek, N.N., Jacques-Palaz, K., Mee, M., Mariscalco, M., Dunny, G.M., Snuggs, M., Winkle, W.B. Van, & Simon, S.I. (1999). *Enterococcus faecalis* bearing aggregation substance is resistant to killing by human neutrophils despite phagocytosis and neutrophil activation. *Infection and Immunity, 67*, 6067-6075.

Rich, R.L., Kreikemeyer, B., Owens, R.T., LaBrenz, S., Narayana, S.V.L., Weinstock, G.M., Murray, B.E., & Höök, M. (1999). Ace is a collagen-binding MSCRAMM from *Enterococcus faecalis*. *Journal of Biological Chemistry*, *274*, 26939–26945.

Sandoe, J.A.T., Witherden, I.R., Cove, J.H., Heritage, J., & Wilcox, M.H. (2003). Correlation between enterococcal biofilm formation *in vitro* and medical-device-related infection potential *in vivo*. *Journal of Medical Microbiology, 52*, 547-550.

Sartingen, S., Rozdzinski, E., Muscholl-Silberhorn, A., & Marre, R. (2000). Aggregation substance increases adherence and internalization, but not translocation, of *Enterococcus faecalis* through different intestinal epithelial cells *in vitro*. *Infection and Immunity, 68*, 6044-6047.

Schwarz, F.V., Perreten, V., & Teuber, M. (2001). Sequence of the 5kb conjugative multiresistance plasmid pRE25 from *Enterococcus faecalis* RE25. *Plasmid, 46*, 170-187.

Shankar, N., Baghdayan, A.S., & Gilmore, M.S. (2002). Modulation of virulence within a pathogenicity island in Vancomycinresistant *Enterococcus faecalis*. *Nature, 417*, 746-50.

Shankar, V., Baghdayan, A.S., Huycke, M.M., Lindahl, G., & Gilmore, M.S. (1999). Infection-derived *Enterococcus faecalis* strains are enriched in *esp*, a gene encoding a novel surface protein. *Infection and Immunity, 67*, 193-200.

Shankar, V., Lockatell, C.V., Bagdayan, A.S., Dranchenberg, C., Gilmore, M.S., & Johnson, D.E. (2001). Role of *Enterococcus faecalis* surface protein esp in the pathogenesis of ascending urinary tract infection. *Infection and Immunity, 69*, 4366-4372.

Stalhammar-Carlemalm, M., Areschoug, T., Larsson, C., & Lindahl, G. (1999). The R28 protein of *Streptococcus pyogenes* is related to several group B streptococcal surface proteins, confers protective immunity and promotes binding to human epithelial cells. *Molecular Microbiology, 33*, 208–219.

Steck, N., Hoffmann, M., Sava, I.G., Kim, S.C., Hahne, H., Tonkonogy, S.L., Mair, K., Krueger, D., Pruteanu, M., Shanahan, F., Vogelmann, R., Schemann, M, Kruster, B., Sartor, R.B., & Haller, D. (2011). *Enterococcus faecalis* Metalloprotease Compromises Epithelial Barrier and Contributes to Intestinal Inflammation. *Gastroenterology, 141*, 959–971.

Stiles, M.E., & Holzapfel, W.H. (1997). Lactic acid bacteria of foods and their current taxonomy. *International Journal of Food Microbiology, 36*, 1-29.

Su, Y.A., Sulavik, M.C., He, P., Mäkinen, K.K., Mäkinen, P., Fiedler, S., Wirth, R., & Clewell, D.B. (1991). Nucleotide sequence of the gelatinase gene (*gelE*) from *Enterococcus faecalis* subsp. *liquefaciens*. *Infection and Immunity, 59*, 415-420.

Suzuki, T., Wada, T., Kozai, S., Ike, Y., Gilmore, M.S., & Ohashi, Y. (2008). Contribution of secreted proteases to thepathogenesis of postoperative *Enterococcus faecalis* endophthalmitis. *Journal of Cataract & Refractive Surgery, 34*, 1776–1784.

Tendolkar, P.M., Baghdayan, A.S., & Shankar, N. (2005). The N-terminal domain of enterococcal surface protein, Esp, is sufficient for Esp-mediated biofilm enhancement in *Enterococcus faecalis*. *Journal of Bacteriology, 187*, 6213–6222.

Tendolkar, P.M., Baghdayan, A.S., & Shankar, N. (2006) Putative surface proteins encoded within a novel transferable locus confer a high-biofilm phenotype to *Enterococcus faecalis*. *Journal of Bacteriology, 188*, 2063–2072.

Thomas, V.C., Thurlow, L.R., Boyle, D., & Hancock, L.E. (2008). Regulation of autolysis-dependent extracellular DNA release by *Enterococcus faecalis* extracellular proteases influences biofilm development. *Journal of Bacteriology, 190*, 5690–5698.

Thurlow, L.S., Thomas, V.C., Narayanan, S., Olson, S., Fleming, S.D., & Hancock, L.E. (2010). Gelatinase contributes to the pathogenesis of endocarditis caused by *Enterococcus faecalis*. *Infection and Immunity, 78,* 4936–4943.

Toledo-Arana, A., Valle, J., Solano, C., Arrizubieta. M.J., Cucarella, C., Lamata, M., Amorena, B., Leiva, J., Penadés, J. R., & Lasa, I. (2001). The enterococcal surface protein, Esp, is involved in *Enterococcus faecalis* biofilm formation. *Applied and Environmental Microbiology, 67,* 4538-4545.

Vu, J., & Carvalho, J. (2011). *Enterococcus*: review of its physiology, pathogenesis, diseases and the challenges it poses for clinical microbiology. *Frontiers of Biology, 6,* 357–366.

Wang, X., & Huycke, M.M. (2007). Extracellular Superoxide Production by *Enterococcus faecalis* Promotes Chromosomal Instability in Mammalian Cells. *Gastroenterology, 132,* 551–561.

Waar, K., Muscholl-Silberhorn, A.B., Willems, R.J., Slooff, M.J., Harmsen, H.J., & Degener, J.E. (2002). Genogrouping and incidence of virulence factors of *Enterococcus faecalis* in liver transplant patients differ from blood culture and fecal isolates. *Journal of Infectious Diseases, 185,* 1121-1127.

Xu, Y., Murray, B. E., Weinstock, G.M. (1998). A cluster of genes involved in polysaccharide biosynthesis from *Enterococcus faecalis* OG1RF. *Infection and Immunity, 66,* 4313-4323.

Zhanel, G.G., Hoban, D.J., & Karlowsky, J.A. (2001). Nitrofurantoin is active against Vancomycin-Resistant Enterococci. *Antimicrobial agents and chemotherapy, 45,* 324-326.

In: *Enterococcus faecalis*
Editor: Henry L. Mack

ISBN: 978-1-63321-049-3
© 2014 Nova Science Publishers, Inc.

Chapter 2

NATURALLY-DERIVED MOLECULES AS A STRATEGY FOR COUNTERING *E. FAECALIS* INFECTION

David M. Pereira[1,2*]

[1]3B's Research Group - Biomaterials, Biodegradables and Biomimetics, University of Minho, Headquarters of the European Institute of Excellence on Tissue Engineering and Regenerative Medicine, Guimarães, Portugal
[2]ICVS/3B's - PT Government Associate Laboratory, Braga/Guimaraes, Portugal

ABSTRACT

Enterococcus faecalis is a gram-positive bacteria that, while a frequent gut commensal, is one of the leading causes of nosocomial infections, which comprise urinary tract infections, endocarditis, bacteremia and meningitis. An important clinical feature of this species is the resistance to a wide range of antimicrobial agents, as demonstrated in clinical, food and water isolates. It not only contains several natural antibiotic resistances, but it is also capable of acquiring new ones as a result of mutations or by acquisition of new genes. Thus, there is a continuous need to search for new drugs that may be used against *E. faecalis*. Some naturally occurring chemical compounds have played a central role in antibiotic drug discovery, with a very significant percentage of clinically proven drugs being derived from natural products. Recently, studies have been reporting that even commonly used herbs, fruits or vegetables, may contain molecules that could constitute potential new treatment against several bacterial infections, including multi-drug resistant bacteria. The present chapter focuses on the most recent published reports on naturally-derived antimicrobial molecules effective against *E. faecalis*. When available, the molecular mechanism of action will also be addressed.

Keywords: *E. faecalis*; Natural products; Fungi; Plants

---

[*] Corresponding author: Email: david.am.pereira@gmail.com.

FIGHTING *E. FAECALIS*

Enterococci are well known organisms in the context of human and animal health, mainly due to their role in diseases such as endocarditis which is, without a doubt, the most relevant pathology if we consider its high mortality rate without effective antimicrobial therapy (Moellering Jr 1992).

Enterococci are Gram-positive, facultative anaerobic *cocci* that can display chains which in turn may present distinct lengths. From a metabolic point of view, these species are known for their ability to survive under harsh conditions, from temperatures ranging from 10 to 45 °C to high salt concentrations (Sakagami, Iinuma et al. 2005). This resistance seems, at least in part, a consequence of their remarkable malleable genome (Paulsen, Banerjei et al. 2003, Arias, Panesso et al. 2011), which may provide evolutionary advantage when challenged with such disadvantageous conditions.

Despite of their pathogenic potential, *enterococci* frequently display low levels of virulence, as evidenced by their presence as natural colonizers of the gastrointestinal (GI) tract in most humans and animals and by the fact that they have been used safely for decades as probiotics in humans and farm animals. Both microbial and host factors can contribute to the conversion of a second-rate pathogen into a first-rate clinical problem. For the *enterococci*, such factors appear to include their inherent ability to resist antimicrobial agents (for example, clindamycin, cephalosporins and aminoglycosides) and their capacity to acquire and disseminate determinants of antibiotic resistance. Several phenotypes of vancomycin resistance are known, however VanA and Van B, encoded by the gene clusters *vanA* and *VanB*, respectively, are amongst the most representative (Arthur, Reynolds et al. 1996).

Moreover, the increasing number of patients who are hospitalized in critical care units and are immunosuppressed, mechanically compromised (by medical devices) and receiving combined antimicrobial agents results in a selective pressure that results in the ability of multidrug-resistant organisms, such as *enterococci*, to cause disease. In this Chapter, some recently described natural molecules that have shown to be effective against *E. faecalis* will be discussed. Further details on *E. faecalis* biology and mechanisms of drug resistance can be found in other Chapters of this Book.

THE PERKS OF USING NATURAL PRODUCTS AS ANTIMICROBIALS

Even nowadays, Nature remains one of the most interesting sources of molecules with potential use in human pharmacotherapy, antibiotics included (Saleem, Nazir et al. 2010). Although plants remain one of the most widely studied sources of compounds (Newman and Cragg 2007, Pereira, Valentao et al. 2012), the Sea has been increasingly regarded as highly promising source of molecules with remarkable chemical diversity and, hence, biological properties (Pereira, Correia-da-Silva et al. 2011, Pereira, Valentão et al. 2013).

The widespread use of natural products as a source of bioactive molecules is not a coincidence. In fact, the rich chemistry of most natural products is one of the most striking factors for their biological properties, with their complex chirality rendering them excellent compounds to bind to complex proteins and other three-dimensional biological targets

(Kingston 2009). This property is further conjugated with other features, namely complex ring systems and the number of heteroatoms and aromatic rings. From the conjugation of all these characteristics, the concept of "privileged structures" arose, meaning that some molecular scaffolds are able to accommodate several pharmacophores, thus displaying multiple biological activities (Costantino and Barlocco 2006).

Nevertheless, it is undeniable that in the last 15 years there has been a decline in the approvals of Nature-derived molecules and many pharmaceutical industries have terminated their natural drug discovery pipeline (Li and Vederas 2009). However, more than a threat this constitutes an opportunity, as the paradigm of drug discovery from natural products is changing with new hotspots of bioactivity under study and techniques for structure elucidation and target characterization.

In the next pages we will briefly discuss some natural molecules that have been shown to be effective antibacterials against *E. faecalis*.

PLANT-DERIVED ANTIMICROBIALS

Plans produce a near limitless amount of chemical compounds, many of which with marked biological properties, in particular those derived from the secondary metabolism. These compounds are frequently synthesized as a defense mechanism against insults such as microbial and insect attack.

Nowadays, among the 25-50% of plant-derived molecules marketed in the USA, only a vestigial amount is intended to be used as antimicrobials. In fact, from an historical point of view, plants are not among the most relevant origins for antibiotics, a role played by fungal and bacterial sources instead (Clark 1996).

α-Mangostin

Α-Mangostin (Figure 1, Table 1) and β-mangostin are xanthones that can be obtained from the mangosteen fruit, a widely consumed product due to its much appreciated organolpetic characteristics, or from the bark of related species such as *G. mangostana*.

These compounds were evaluated against 5 strains of vancomycin-resistant *enterococci* (VRE) and 3 strains of vancomycin-sensitive Enterococcus (VSE) (Sakagami, Iinuma et al. 2005). Overall, α-mangostin, but not its β homologue, was shown to be a very promising antibiotic as it was active against all VRE with minimum inhibitory concentration (MIC) in the range 3.13-6.25 µg/ml. Marked activity was also found for several strains of MRSA (Sakagami, Iinuma et al. 2005).

The ability of mangostin to act synergically with clinically-relevant antibiotics was also investigated. By employing the fractional inhibitory concentration (FIC) index calculations, the authors proved that α-mangostin displayed relevant synergism with gentamycin, while partial synergism with ampicillin, minocycline, fosfomycin and vancomycin against VRE was reported (Sakagami, Iinuma et al. 2005).

Pentacyclic triterpenes

Ursolic and oleanolic acids (Figure 1, Table 1) are pentacyclic triterpenes (C_{30} compounds that are formed from the C_5 building block isoprene) quite widespread in Nature, were they occur in several medicinal and dietary-relevant plants.

These two compounds were evaluated for their antibacterial activity, *E. faecalis* included (Fontanay, Grare et al. 2008). Overall, these triterpenes displayed differential activity, with ursolic acid causing a MIC of 4 µg/ml and oleanolic acid 8 µg/ml.

Interestingly, both compounds were ineffective against Gram-negative bacteria (Fontanay, Grare et al. 2008), which may suggest a mechanism of action that exploits the differences in these groups' cell walls.

Rhodomyrtone

Rhodomyrtone (Figure 1, Table 1) is a promising acylphloroglucinol (the largest group of the phloroglucinol class) isolated for the first time from the leaves of *Rhodomyrtus tomentosa* (Salni, Sargent et al. 2002). In subsequent studies, this molecule was identified in a bioactivity-guided assay, where it was assessed for its antibacterial activity against Gram-positive bacteria including *Bacillus cereus*, *Bacillus subtilis*, *E. faecalis*, *Staphylococcus aureus*, MRSA, *Staphylococcus epidermidis*, *Streptococcus gordonii*, *Streptococcus mutans*, *Streptococcus pneumoniae*, *Streptococcus pyogenes*, and *Streptococcus salivarius* (Limsuwan, Trip et al. 2009). In the case of *E. faecalis*, the MIC found was 1.56 µg/ml. The compound was also markedly effective against several clinically-relevant bacteria, in particular MRSA, with a MIC of 0.39 µg/ml. In the following studies, the compound was tested against several hospital-acquired antibiotic-resistant VRE strains (Leejae, Taylor et al. 2013). In several VRE clinical isolates, namely VRE-2 VRE-3 VRE-7 VRE-8, rhodomyrtone was very effective, with MIC ranging from 1-2 µg/ml. In the particular case of *E. faecalis* ATCC 29212, MIC was 2 µg/ml, the same value found for vancomycin (Leejae, Taylor et al. 2013).

FUNGI-DERIVED ANTIMICROBIALS

An approximate value of 25% of all biologically active natural products are believed to be of fungal origin, a number that is, most likely, underestimated due to the fact that we can, nowadays, culture around 7000 species, with 1.5 million species being believed to exist (Brady, Singh et al. 2000).

When we consider Nature-derived antimicrobials, fungi are, without a doubt, one of the most prolific sources of drugs, a consequence of a process of co-evolution in which these organisms are required to cope and sometimes compete with bacteria. In fact, if we consider the number from the 2000's, around 2-3 antibiotics of microbial origin enter the market every year (Clark 1996).

Table 1. Chemical class, origin and references regarding the molecules discussed in this chapter

Compound	Class	Species	Reference
α-Mangostin	Xanthone	Several	(Sakagami, Iinuma et al. 2005)
Ursolic acid	Pentacyclic triterpene	Several	(Fontanay, Grare et al. 2008)
Oleanolic acid	Pentacyclic triterpene	Several	(Fontanay, Grare et al. 2008)
Guanacastepene	Neodolastane diterpenes	Endophytic fungus CR115	(Brady, Singh et al. 2000, Brady, Bondi et al. 2001).
Rhodomyrtone	Phloroglucinol	*Rhodomyrtus tomentosa*	(Limsuwan, Trip et al. 2009)
Cytosporone D	Trihydroxybenzene lactone	*Cytospora* sp.	(Brady, Wagenaar et al. 2000)
Cytosporone E	Trihydroxybenzene lactone	*Diaporthe* sp.	(Brady, Wagenaar et al. 2000)
Avrainvillamide	Alkaloid	*Aspergillus ochraceus*	(Sugie, Hirai et al. 2001)
Pseudopterosin P	Diterpene	*Pseudopterogorgia elisabethae*	(Ata, Win et al. 2004)
pseudopterosin Q	Diterpene	*Pseudopterogorgia elisabethae*	(Ata, Win et al. 2004)
Usnic acid	Dibenzofuran derivative	Several lichens	(Lauterwein, 1995)

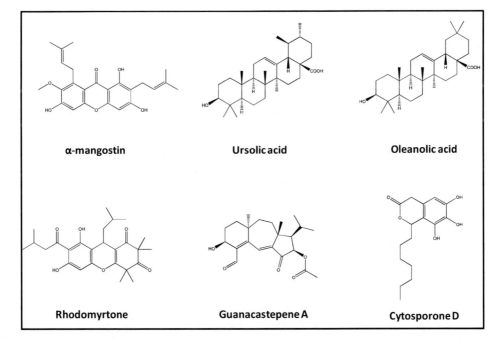

Figure 1. Structures of some compounds discussed in this Chapter.

Guanacastepene

Guanacastepene A (Figure 1, Table 1) was the first of a series of neodolastane diterpenes isolated for the first time from the endophytic fungus CR115 collected in the tree *Daphnopsis Americana* (Brady, Singh et al. 2000, Brady, Bondi et al. 2001).

Screening tests using agar diffusion assays showed that this molecule displayed antibiotic activity against both methicillin-sensitive and -resistant *S. aureus* and vancomycin-resistant *E. faecalis* (VREF). As expected, vancomycin was innefective against VREF, however 100 μg of guanacastepene elicited a 9 mm zone of growth inhibition and, in the case of MRSA, 17 mm (Brady, Singh et al. 2000).

Cytosporones

Several octaketides were isolated from endophytic fungus collected from the tissues of *Conocarpus erecta* and *Forsteronia spicata* plants (Brady, Wagenaar et al. 2000). These compounds are trihydroxybenzene lactones and were generally named as cytosporones, comprising both compounds displaying antibacterial activity and those devoid of it. In particular, cytosporones D (Figure 1, Table 1) and E (Figure 2, Table 1) revealed an MIC against *S. aureus, E. faecalis, Escherichia coli* and the fungus *Candida albicans* of 8, 8, 64, and 4 μg/mL, respectively. It is interesting to realize that the trihydroxybenzene moiety seems to be critical for the antibacterial activity, as showed by the fact that cytosprone C, a closely related compound without such moiety was devoid of antibiotic activity (Brady, Wagenaar et al. 2000).

Avrainvillamide

Yet another fungal-derived antibiotic, CJ-17,665, was isolated from the fermentation brooth of *Aspergillus ochraceus* (Sugie, Hirai et al. 2001). This is a very interesting molecule as it exhibits some traits that are uncommon in natural products, namely a diketopiperazine and an indole N-oxide moieties. This compound was shown to be effective against *E. faecalis*, with a MIC of 25 μg/mL. In subsequent studies, this compound was named avrainvillamide (Figure 2, Table 1) and, despite its challenging structure, synthesis of this molecule has already been reported (Baran, Guerrero et al. 2005, Baran, Hafensteiner et al. 2006)

Lynamicins

Lynamicin B (Figure 2, Table 1) is a chlorinated bisindole pyrrole isolated from a recently described novel marine actinomycete, NPS12745 (McArthur, Mitchell et al. 2008). Other compounds of the same family have been obtained from the same organism, although their activity against *E. faecalis* is lower. Against a strain of VSE, the MIC of lynamicin B was 1.5 μg/mL, while other compounds from the same series, lynamicin A, C and D displayed MICs of 8, 2.5 and 8 and <24 µg/mL, respectively (McArthur, Mitchell et al. 2008).

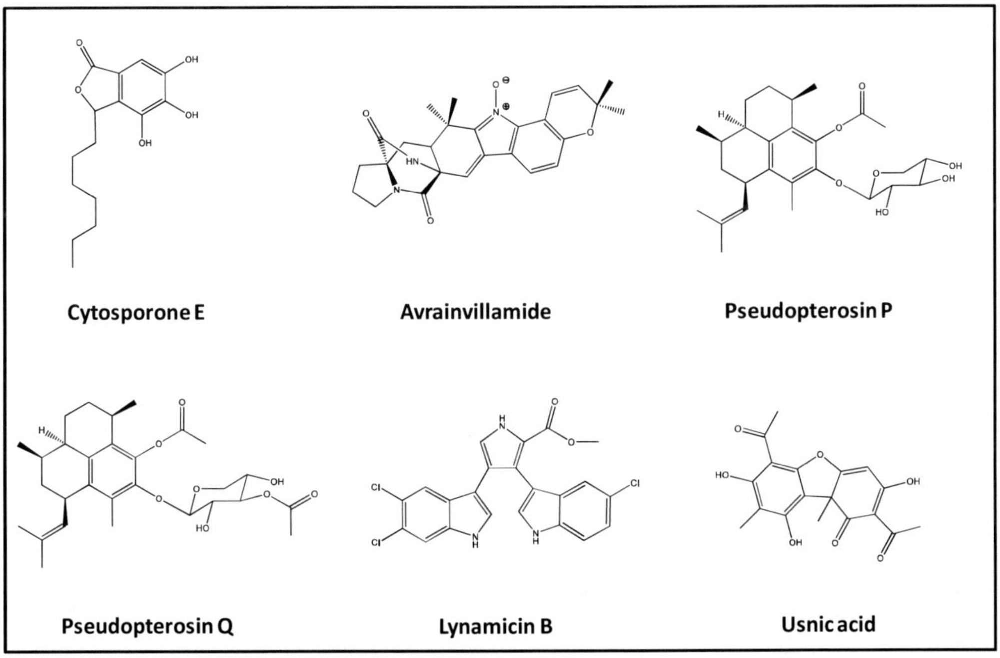

Figure 2. Structures of some compounds discussed in this Chapter.

OTHER SOURCES

Pseudopterosins

The caribean octacoral *Pseudopterogorgia elisabethae* has been the source of the diterpenes elisabethin E, elisabethin F, pseudopterosin P and pseudopterosin Q (Ata, Win et al. 2004). These molecules were evaluated for their antibacterial capacity using the agar disk diffusion assay, at a concentration of 25 µg/ml, against a panel of several bacteria, *E. faecalis* included.

While elisabethin E and F were devoid of antibacterial activity, pseudopterosin P and Q (Figure 2, Table 1) selectively inhibited the growth of the Gram-positive bacteria *S. pyogenes, S. aureus* and *E. faecalis*, the diameter of inhibition being 8 mm in the case of the latter. In Gram-negative bacteria, *E. coli* and *P. aeruginosa*, no activity was found.

Other previously described compounds, including pseudopterosins A- L, were also tested, with similar values being found (Ata, Win et al. 2004).

Usnic Acid

Lichens are formed from the symbiosis of fungi and algae/cyanobacteria. Several studies show that, from a chemical point of view, the most common molecules involved in lichen metabolism are depsides, depsidones and dibenzofurans, with the dibenzofuran derivative usnic acid (Figure 2, Table 1) being one of the most widely studied metabolite (Cocchietto, Skert et al. 2002, Ingolfsdottir 2002).

In a study aiming to evaluate its antimicrobial activity against *E. faecalis*, marked differences between the activity of (–) and (+) usnic acids were found, the latter displaying a MIC of 4 µg/ml, half of that found for its (-) homologue (Lauterwein, Oethinger et al. 1995).

FUTURE TRENDS

During most of human History, natural products have been the sole sources of medicines to be used in human pharmacotherapy. Nowadays, although alternative sources are available, Nature is still a prolific source of new chemical entities (NCE), many of which with marked biological activities. In the near future, the role of Nature as a source of medicines is likely to continue, although a shift in paradigm, which is already in course, is expected to take place. This paradigm relies in the search for new molecules in "non-conventional" sources of natural products, namely marine microorganisms and macro-invertebrates as well as extremophiles.

Still, although Nature is a reliable and promising source of NCE, we must be careful with the interpretation of many of the results available nowadays. Ideally, in drug discovery from Natural sources crude extracts are initially screened and, in the case of bioactivity, the compounds responsible should be identified by bio-guided isolation. Frequently, the activity displayed by extracts cannot be fully reproduced with pure compounds due to the occurrence of phenomenon of synergism or antagonism. That said, a great number of works still test raw extracts for antimicrobial activity without further information regarding the compounds responsible being provided.

Another topic that will require further attention from researchers is related with the lack of information regarding the mechanism of action of many antimicrobial molecules. As so, most of the papers available nowadays do not provide enough detail regarding the molecular mechanism of action of such molecules. Due to the widespread distribution of antibiotic resistances across several bacterial species, it is particularly important to know the molecular targets of drugs. This information could, eventually, lead to the discovery of novel classes of antibiotics that modulate targets that are yet to be involved in resistance.

On another topic, in the particular case of drug discovery in the area of antibiotics, much importance must be given to the exact strain of microorganism used and conditions of the assay. As so, in many works it is not possible to be sure the precise strain of the bacteria used, thus hindering direct comparison between the results of distinct groups. This information is particularly important in the case of clinical isolates.

That said, Nature is expected to remain an unavoidable source of molecules for antimicrobial chemotherapy by providing new molecules to counter infections and, hopefully, by empowering us with the ability to overcome antimicrobial resistance.

REFERENCES

Arias, C. A., et al. (2011). "Genetic basis for in vivo daptomycin resistance in *Enterococci.*" *New England Journal of Medicine, 365*(10), 892-900.

Arthur, M., et al. (1996). "Glycopeptide resistance in *Enterococci.*" *Trends in Microbiology, 4*(10), 401-407.

Ata, A., et al. (2004). "New antibacterial diterpenes from *Pseudopterogorgia elisabethae.*" *Helvetica chimica acta, 87*(5), 1090-1098.

Baran, P. S., et al. (2005). "Total synthesis of avrainvillamide (CJ-17,665) and stephacidin B." *Angewandte Chemie, 117*(25), 3960-3963.

Baran, P. S., et al. (2006). "Enantioselective total synthesis of avrainvillamide and the stephacidins." *Journal of the American Chemical Society, 128*(26), 8678-8693.

Brady, S. F., et al. (2001). "The guanacastepenes: a highly diverse family of secondary metabolites produced by an endophytic fungus." *Journal of the American Chemical Society, 123*(40), 9900-9901.

Brady, S. F., et al. (2000). "Guanacastepene, a fungal-derived diterpene antibiotic with a new carbon skeleton." *Journal of the American Chemical Society, 122*(9), 2116-2117.

Brady, S. F., et al. (2000). "The cytosporones, new octaketide antibiotics isolated from an endophytic fungus." *Organic letters, 2*(25), 4043-4046.

Clark, A. M. (1996). "Natural products as a resource for new drugs." *Pharmaceutical Research, 13*(8), 1133-1141.

Cocchietto, M., et al. (2002). "A review on usnic acid, an interesting natural compound." *Naturwissenschaften, 89*(4), 137-146.

Costantino, L. and D. Barlocco (2006). "Privileged structures as leads in medicinal chemistry." *Current Medicinal Chemistry, 13*(1), 65-85.

Fontanay, S., et al. (2008). "Ursolic, oleanolic and betulinic acids: antibacterial spectra and selectivity indexes." *Journal of Ethnopharmacology, 120*(2), 272-276.

Ingolfsdottir, K. (2002). "Usnic acid." *Phytochemistry, 61*(7), 729-736.

Kingston, D. G. I. (2009). "Tubulin-interactive natural products as anticancer agents." *Journal of Natural Products, 72*(3), 507-515.

Lauterwein, M., et al. (1995). "In vitro activities of the lichen secondary metabolites vulpinic acid,(+)-usnic acid, and (-)-usnic acid against aerobic and anaerobic microorganisms." *Antimicrobial Agents and Chemotherapy, 39*(11), 2541-2543.

Leejae, S., et al. (2013). "Antibacterial mechanisms of rhodomyrtone against important hospital-acquired antibiotic-resistant pathogenic bacteria." *Journal of Medical Microbiology, 62*(Pt 1), 78-85.

Li, J. W.-H. and J. C. Vederas (2009). "Drug discovery and natural products: end of an era or an endless frontier?" *Science, 325*(5937), 161-165.

Limsuwan, S., et al. (2009). "Rhodomyrtone: A new candidate as natural antibacterial drug from *Rhodomyrtus tomentosa.*" *Phytomedicine, 16*(6), 645-651.

M Pereira, D., et al. (2012). "Plant secondary metabolites in cancer chemotherapy: where are we?" *Current Pharmaceutical Biotechnology, 13*(5), 632-650.

McArthur, K. A., et al. (2008). "Lynamicins A− E, Chlorinated bisindole pyrrole antibiotics from a novel marine actinomycete†." *Journal of natural products, 71*(10), 1732-1737.

Moellering Jr, R. C. (1992). "Emergence of *Enterococcus* as a significant pathogen." *Clinical Infectious Diseases,*: 1173-1176.

Newman, D. J. and G. M. Cragg (2007). "Natural products as sources of new drugs over the Last 25 Years⊥." *Journal of Natural Products*, *70*(3), 461-477.

Paulsen, I., et al. (2003). "Role of mobile DNA in the evolution of vancomycin-resistant *Enterococcus faecalis*." *Science*, *299*(5615), 2071-2074.

Pereira, D. M., et al. (2011). Marine metabolomics in cancer chemotherapy. *OMICS - Biomedical Perspective and Applications,*. D. Barh, CRC Press: 377-398.

Pereira, D. M., et al. (2013). Lessons from the Sea: distribution, SAR and molecular mechanisms of anti-inflammatory drugs from marine organisms. *Studies in Natural Products Chemistry*. Atta-ur-Rahman. The Netherlands, Elsevier Science Publishers. *39*, 205-228.

Sakagami, Y., et al. (2005). "Antibacterial activity of α-mangostin against vancomycin resistant Enterococci (VRE) and synergism with antibiotics." *Phytomedicine*, *12*(3), 203-208.

Saleem, M., et al. (2010). "Antimicrobial natural products: an update on future antibiotic drug candidates." *Natural Product Reports*, *27*(2), 238-254.

Salni, D., et al. (2002). "Rhodomyrtone, an antibotic from *Rhodomyrtus tomentosa*." *Australian Journal of Chemistry*, *55*(3), 229-232.

Sugie, Y., et al. (2001). "A new antibiotic CJ-17,665 from *Aspergillus ochraceus*." *Journal of Antibiotics*, *54*(11), 911-916.

In: *Enterococcus faecalis*
Editor: Henry L. Mack

ISBN: 978-1-63321-049-3
© 2014 Nova Science Publishers, Inc.

Chapter 3

ENTEROCOCCUS FAECALIS: ROLE IN NOSOCOMIAL INFECTION, RESISTANCE TRAITS AND MOLECULAR EPIDEMIOLOGY

Juliana Caierão[*]
Federal University of Health Science of Porto Alegre (UFCSPA)

ABSTRACT

Enterococci are recognized by their physiological versatility, which is responsible for the ubiquitous occurrence of these microorganisms. Because of this extraordinary capacity to survive under unfavorable conditions, they can persist in nosocomial environment for long periods, which may represent the source of exogenous enterococcal infections.

In humans, they compose genitourinary, oral and especially gastrointestinal microbiota. *Enterococcus faecalis* is, by far, the major species in both, colonized or infected patients. Although *Enterococcus faecium* is well-recognized by its resistance, *E. faecalis* consistently presents a more robust virulence arsenal. Its virulence has been defined as multifactorial, with participation of many different molecules and features, especially adhesins and the capacity of biofilm production.

Until the 70s, their pathogenic role had been neglected. However, since that period, they have been recognized as one of the leading cause of opportunistic infections in nosocomial setting, especially affecting immunosupressed, elderly or long-term hospitalized patients. The major clinical syndromes related to enterococci are bacteremia, endocarditis and urinary tract infections. Besides, they may be frequently associated to biliary, abdominal and wound infections.

The acceptance of this pathogenicity was coincident, and possibly related with the increase use of broad-spectrum antimicrobial agents, such as third-generation cephalosporins, to whom enterococci are intrinsically resistant. Indeed, this intrinsic resistance is extended to the majority of antimicrobials commonly used to treat gram-positive cocci infections. Therefore, this feature makes them much more adapted to the nosocomial environment than other bacterial genus.

---

[*] E-mail: julianaca@ufcspa.edu.br.

Along with their intrinsic resistance characteristics, enterococci present an extraordinary capacity to acquire mobile genetic elements, carrying resistance genes to different classes of antimicrobials, including: chloramphenicol, tetracyclines, macrolides, glycopeptides and high levels of aminoglycoside. Therefore, efficient antimicrobials are scarce, leading to difficulties in treatment of enterococcal infections. The occurrence and dissemination of multidrug-resistant strains is well-recognized and represents a challenge to medical and infection control staff.

The most impactful phenotype is the Vancomycin-Resistant Enterococci (VRE), commonly related to multidrug-resistant isolates. Despite the introduction of new drugs active against VRE, resistance to them, including linezolid, tigecycline and daptomycin, have been described around the world. VRE genotype may be related to nine different *van* genes, located in mobile genetic elements. The most relevant genotype is, by far, *vanA*-VRE, which may present a clonal or heterogeneous dissemination, although the former seems to be more frequent and is associated with exogenous acquisition through healthcare staff. Long-term hospitalization, stay in intensive therapy unit and previous usage of antimicrobials have been associated to VRE acquisition.

VRE is endemic in many regions of the world and management of outbreaks requires strategies to avoid new cases and to reduce transmission rates, which includes isolation of infected or colonized patients. This management is difficult because, once established in determined nosocomial setting, VRE is hard to eradicate.

In conclusion, enterococci is a challenging opportunistic pathogen in nosocomial settings, especially because of its resistance traits, ability to survive for long periods in the environment, difficulty to eradicate and control its dissemination and the multifactorial virulence, which can cause life threatening infections in specific and severely ill patients.

Because of their physiological characteristics, such as growth in a wide range of temperature and pH and in a high salt concentration, *Enterococcus* spp. are ubiquitous, being largely distributed in nature (soil, plants and water). As they are nutritionally undemanding and versatile, they can survive in hospital environment for more than five weeks. In animals and humans, they are part of normal microbiota of oral cavity and genitourinary tract (especially among females). In gastrointestinal tract, although it represents a small percentage of the total microbiota, enterococci are the most common gram-positive cocci normally found.

Historically, enterococci were considered merely commensals or contaminants of clinical cultures, based on its low virulence. Indeed, they have been used as safely probiotics for decades. However, since the middle of 70s, some species became recognized as important opportunistic nosocomial pathogens, affecting, predominately, immunocompromised patients, elderly, patients with injuries in the normal defense barriers or the ones hospitalized for long periods.

This acceptance of the pathogenic role of *Enterococcus* spp. was coincident, and possibly related, to the increase use of broad spectrum antimicrobials, such as third generation cephalosporins, to which enterococci are intrinsically resistant, enable them adaptive advantages in the hospital environment with increasing importance as a nosocomial opportunistic pathogen.

Indeed, *Enterococcus* spp. are considered the second leading cause (after staphylococci) of nosocomial infections in the United States (US), including catheter-associated bacteremias, urinary tract and skin and soft-tissue infections. In bacteremias, *Enterococcus* spp. are among the three most common etiological agent in the US and in many countries of Europe.

Enterococcal bacteremias are, in general, severe and related to worrisome mortality rates, which has been reported to be as high as 51%.

The role of enterococci in infective endocarditis is recognized since 1899 and, nowadays, they are among the most common etiological agents causing endocarditis in the US. This genus is particularly associated to native valves, where around 20% of all cases are associated to them. Most studies demonstrate that in infective process of prosthetic valve, participation of enterococci as etiological agent is less frequent (near 8%), although some authors evidence higher values (around 20%). Indeed, a large international prospective cohort (including 2781 cases of endocarditis) rank *Enterococcus* spp. as the third most frequent cause of both native and prosthetic valve, after *Staphylococcus* spp. and *Streptococcus* spp. Recent studies suggest that the frequency of enterococcal endocarditis is increasing, especially among health-care associated endocarditis, where enterococci are considered the second most frequent etiological agents, suppressed only by staphylococci.

Enterococcal bacteremia may result in infective endocarditis, but the frequency that it occurs varies widely in different publications. There are some well-established risk factors for it, which include a history of pre-existent valvular heart disease, presence of prosthetic valve and infection with the species *E. faecalis*. Some people are more susceptible to develop enterococcal endocarditis, such as older patients with underlying diseases and previous valvular damage or a prosthetic valve. The main complications of enterococcal endocarditis is heart failure, which occurs in almost half of the patients and has an important impact on outcome.

The management of enterococcal endocarditis has long been recognized as a challenging clinical problem. One of the reasons for it is that endovascular infections, such as endocarditis, are entities in which bactericidal therapy appears to be of paramount importance for eradication of infection organisms and clinical cure. Despite the improvement in therapeutical schemes, and also in diagnosis, mortality due to enterococcal endocarditis has not changed significantly over the last years. Even though, this mortality is lower than in other causes, especially when compared with *Staphylococcus aureus*.

Besides those classical clinical infections, enterococci also play important role in surgical wound infections, which are, in general, polymicrobial, and urinary tract infections (UTI). The incidence of UTI caused by enterococci varies from 4 to around 20%, depending on the origin (nosocomial or community-acquired infections) and the patient's characteristics, such as age, presence of catheter, underlying disease and gender. Indeed, enterococci are more prevalent among female UTI compared to male infections. Respiratory tract infections, nervous system infection, as well as otitis, endophthalmitis, septic arthritis, among others may rarely occur.

Historically, the most frequently recovered species from human infections are, by far, *Enterococcus faecalis* (approximately 90%), followed by *Enterococcus faecium* (5-10%). However, more recent studies strongly suggest that the proportion of infections caused by *E.faecium* has increasing compared to *E. faecalis*, possibly due to some specific clonally distributed *E.faecium* lineages (markedly the clonal complex 17, CC17) highly adapted to hospital environment, which include specific virulence and resistance traits.

Even though, although less frequent or even rare, many other species, such as *Enterococcus gallinarum*, *Enterococcus casseliflavus*, *Enterococcus avium*, *Enterococcus mundtii*, *Enterococcus durans*, *Enterococcus raffinosus*, among others have been recovered from human sources. The increase in prevalence of more rare species is related (i) to the fact that patients are becoming more compromised (due to underlying diseases, immunological disorders or surgical procedures) leading them more susceptible to opportunistic infections, and (ii) to advances in identification methodologies. In this context, identification of enterococcal species are a subject of clinical concern, once there are differences in pathogenic potential and antimicrobial susceptibility patterns.

The genus *Enterococcus* spp. present relatively low virulence if compared to other gram-positive cocci, such as *Staphylococcus* spp. and *Streptococcus* spp. Virulence in enterococci is considered multifactorial, which leads to difficulties to define the real role of each component solely. Its virulence is mostly associated to adherence molecules and *E. faecalis* is, by far, the most virulent species within the genus.

Several proteins that are secreted into the extracellular medium have been implicated in enterococcal virulence. Cytolysin (Cyl) is a hemolytic toxin produced by 30% of *E. faecalis* strains and is encoded on pheromone-responsive plasmids or pathogenicity islands. Cyl is secreted extracellularly as two structural subunits (CylL-L and CylL-S) and then proteolytically activated. Cyl can lyse red blood cells from humans, horses and rabbits, but not sheep or cows, and can also lyse some human white blood cells. *E. faecalis* strains expressing *cyl* are more virulent in various animal models than isogenic strains without *cyl*.

Proteases are also involved in enterococcal virulence and the most important are gelatinase (GelE) and the extracellular serine proteinase (SprE). GelE seems to mediate virulence through effects such as degradation of host tissues and modulation of the host immune response. It has an important role in clearing misfolded proteins and participates in the activation of autolysin, a peptidoglycan-degrading enzyme, which leads to the release of extracellular DNA and the formation of a biofilm. The genes encoding these proteases in *E. faecalis* are regulated by the Fsr quorum sensing system, which is homologous to the Agr system of staphylococci (involved in regulating the expression of several virulence factors in *S. aureus*). It has been demonstrated that Fsr system affect the pathogenesis of some *E. faecalis* infections.

Mutants lacking GelE show a marked decrease in biofilm formation, a decrease in translocation across T84 intestinal cells, attenuated virulence in peritonitis, endocarditis, endophthalmitis and reduced adherence to dental roots. It should be noted that Cyl and GelE are seem equally in isolates recovered from clinical infection and in those from stools of healthy individuals, illustrating that various putative enterococcal virulence determinants can be also found in trains colonizing the GI of healthy individuals.

As mentioned above, most enterococcal virulence factors are related to adhesion. The group of aggregation substances (AS) has been extensively studied. The *agg* gene is located in pheromone-responsive plasmids that frequently also carry resistance genes. Expression of AS proteins are induced by pheromone, but can also be induced by a host factor during *in vivo* growth. This virulence factor mediates bacterial aggregation during conjugation, facilitating plasmids exchange. Besides, AS mediates binding to the host epithelium, increase survival within polymorphonuclear neutrophils and facilitate bacterial internalization by different cultured intestinal epithelial cells, indicating that they might be involved in the translocation of *E. faecalis* through the intestinal epithelia, leading to systemic infections.

Experimentally, AS affect pathogenesis of endocarditis by favoring formation of large bacterial aggregates on de cardiac valve. AsaI, AspI and Acs are the best studied AS proteins and show over 90% of amino acid sequence identity. AS proteins contain a N-terminal domain, a variable region, a central domain and two Arg-Gly-Asp (RGD) motifs. This RGD motif probably mediate the interactions with eukaryotic cells and the N-terminal aggregation domain promotes binding to cell wall lipoteichoic acid (LTA).

Apart of the secreted substances, enterococci produce some molecules that remains anchored to the cell. The enterococcal surface protein (Esp) is one of them. It is encoded by *esp* gene, which seems to have been acquired within a pathogenicity island, and contribute to colonization and persistence of *E. faecalis* strains in ascending infections of the urinary tract. In addition, it mediates the interaction with primary surfaces and participates in biofilm formation, which substantially contributes to bacterial survival in biopolymers and may also be involved in antimicrobials resistance. However, some studies failed to demonstrate this close relationship among *esp* and biofilm, as there exist strains able to produce biofilm without *esp* and vice versa.

There is another group of important adhesion molecules, the MSCRAMMs (microbial surface components recognizing adhesive matrix molecules), that act in the early stages of infection, binding to the host extracellular matrix.

Another important adhesion molecule is the collagen adhesin (Ace in *E.faecalis* and Acm in *E. faecium*). These proteins has considerable homology with Cna (Collagen adhesin) of *S. aureus* and has been shown to affect enterococcal pathogenesis in in vitro models. Indeed, *E.faecalis* mutants lacking *ace* gene present an import decrease in pathogenesis in animal models of endocarditis and UTI. This protein acts binding collagen in a specific manner, where Ace appears to embrace the collagen molecule after initial docking.

In *E. faecium* Acm has similar activity. It has been demonstrated that strains recovered from stools of human and animals present a pseudogene of Acm. Therefore, it is reasonable to believe that Acm may have a role in the increase ability of members of the hospital-associated *E. faecium* clade to cause disease.

The enterococcal leucine-rich-repeat-containing protein, ElrA, is another cell wall-associated protein that seems to have an important role in pathogenesis, as mutants lacking the *elra* gene presented attenuated virulence and cannot infect macrophages in animal models of peritonitis.

It is well-known the importance of pili for initial adherence of gram-positive bacteria. *E. faecalis* presents a ubiquitous pili, the endocarditis and biofilm-associated pili (Ebp), which is important for biofilm formation and for the pathogenesis of experimental endocarditis and UTI. The expression of the *ebp* seems also to be regulated by the Fsr system. There is another pili in *E. faecalis*, the Bee (biofilm enhancer in *Enterococcus*), but the occurrence of *bee* gene is infrequent. On the other hand, *E. faecium* frequently harbor four or more putative pili loci, with heterogeneous expression. One of these loci present homology with the *ebp* of *E.faecalis* and seems to affects biofilm formation as well as the pathogenesis of animal models of UTI.

Some strains of *E.faecalis*, especially those of hospital-associated clonal clusters, have a capsular polysaccharide locus (*cps*), which is formed by 8 to 9 genes. Based on the structural differences on the capsular polysaccharide, enterococci can be classified into types C and D. These strains are able to mask LTA, thus conferring resistance to opsonophagocytosis mediated by complement. Isolates that lack *cps* locus are designated as type A and B.

Another cell wall antigen, the enterococcal polysaccharide antigen (Epa), is immunogenic, as it is recognized by sera from most patients with serious infections caused by *E. faecalis*. Disruption of *epa* cluster interfere in biofilm formation and translocation across an enterocyte monolayer, decreasing both occurrence. *E.faecalis* lacking *epa* gene are also more susceptible to polymorphonuclear neutrophil-mediated killing and have its pathogenesis attenuated in animal models of peritonitis.

Megaplasmids, which are 150-250kb in size, are transferable and non-responsive to pheromones, common among *E.faecium* clinical isolates. They have a role in virulence, as the transference of them to commensal strains increase its virulence. However, the precise role of specific genes carried by these plasmids in virulence or colonization remains to be established.

Besides surface molecules, some stress response proteins are also important for enterococcal virulence. *E. faecalis* produces the Gls24 that is responsible for its resistance to bile salts; besides and it has been observed that the absence of *gls*24 attenuate virulence in animal models. Besides, Gls-24 specific immune serum protects mice against a lethal *E. faecalis* challenge in peritonitis model. *E. faecium* harbor similar proteins, also important for virulence.

Protection against reactive oxygen species is an important feature for virulence, once they can protect bacteria in the phagocyte environment. *E. faecalis* produces three peroxidases: NADH peroxidase (Npr), thiol peroxidase (Tpx) and an reductase system: alkyl hydroperoxidase reductase system, Ahp.

Altogether, these above-cited molecules and structures compose the multifactorial enterococcal virulence. However, despite of those important mechanisms, resistance to multiple antimicrobials should also be considered as an important mechanism that indirectly increase enterococcal virulence.

Enterococcus spp. are intrinsically resistant to many antimicrobials habitually used for treatment of gram-positive cocci infections, such as cephalosporins, lincosamides, co-trimoxazole and low-levels of β-lactams, aminoglycosides and glycopeptides (some specific species, i.e., *E.casseliflavus* and *E.gallinarum*). This feature ensures them better adaptive advantages than some other bacteria in hospital environment.

Along with this inherent genus characteristic, enterococci present an extraordinary capacity to acquire mobile genetic elements, which carry resistance genes to multiple antimicrobials, including chloramphenicol, tetracyclines, macrolides, streptogramins and high-levels of glycopeptides, β-lactams and aminoglycosides. These intrinsic plus acquired resistance traits lead to difficulties in treating enterococcal infections and these challenges have been recognized since the 50s to the present days, where spread of multiresistant strains is a subject of major concern to physicians and infection control team.

The great ability to horizontally transfer genetic material is explained, in part, by the occurrence of multiple plasmids and transposons in most of enterococcal clinical strains. These elements may interact with each other or with bacterial chromosome to form composite mobile elements. Most of the acquired resistance genes are located in pheromone-responsive plasmids. These plasmids are found predominantly in *E. faecalis* and its transference occur as follows: pheromones, which are chromosomally encoded lipoprotein fragments released by recipient cells, are sensed by nearby donor cells and stimulate production of AS encoded by plasmids. Aggregation substance interacts with enterococcal binding substance on the surface of the recipient cell and stimulate recipient-donor contact that promotes effective and frequent conjugation. Indeed, these plasmids transmit genetic information in a highly efficient manner between *E. faecalis* strains (10-3/donor cell during 4h mating), but are largely restricted to this species. pRUM plasmids in *E. faecium* are similar to pheromone-responsive plasmidis in

E. faecalis in that they transfer at a high frequency but exhibit a narrow host range. In contrast, broad range plasmids are capable of transferring resistance (or others) genetic information to other gram-positive and even gram-negative species, but at a lower frequency (10-7/donor cell during 4h mating) than pheromone-responsive plasmids.

Among enterococcal species, *E. faecium* has aroused greater interest considering resistance traits because of its peculiar characteristics of resistance to antimicrobials. Complete or relative resistance to β-lactams is a characteristic of the genus. *E. faecalis* is typically 10 to 100 folds less susceptible to penicillin if compared to most *Streptococcus* spp., while *E. faecium* is, at least, 4 to 16 folds less susceptible than *E. faecalis* to the same antimicrobial. The main mechanism responsible to this intrinsic resistance is hyperexpression of penicillin-binding proteins (PBP) with low affinity to β-lactams or the occurrence of mutations in constitutive PBP, which reduce even more their affinity to antimicrobials. *Enterococcus spp.* present, at least, 5 PBP, being the hyperproduction of PBP5 associated to resistance to all β-lactams. Considering mutations, it is well established that the ones occurred in the active site of PBP4 and PBP5 are related to the increase of resistance to penicillin, even if hyperproduction is not observed.

Rarely, β-lactam resistance in enterococci may be related to β-lactamase production. Some strains of *E. faecium* produce a β-lactamase that is identical to the one produced by type A *Staphylococcus* spp. However, in enterococci, production of this enzyme is constitutive and codified by a gene localized in a transferable plasmid along with genes related to high-level resistance to gentamycin. This resistance mechanism do not have wide spread in the genus and has been described predominantly in *E. faecalis*.

Regardless the type of resistance mechanism, ampicillin resistance have disseminated worldwide in the 90s, with some geographical particularities. Indeed, in the US, in 2000, 90% of *E. faecium* strains were resistant to ampicillin. In Europe, frequencies has become similar, but a decade later. Overall, among *E. faecalis*, rates of ampicillin remain low, irrespective of global region.

Another difficulty faced to treat enterococcal infections with β-lactams is the development of tolerance. In these cases, bactericidal effect does not occur anymore and these antimicrobials are only able to inhibit enterococcal growth. This is an acquired characteristics that arises quickly during treatment, which justify the non-use of β-lactams in monotherapy for treatment of severe enterococcal infections. Therefore, they are commonly associated to aminoglycosides to reach a synergistic effect.

Aminoglycosides act primarily interfering in protein synthesis by binding to the 16S rRNA of the 30S ribosomal subunit. Low-level intrinsic resistance is due to a reduction of permeability by inefficient transportation into the cell. Gentamycin Minimal Inhibitory Concentrations (MIC) among enterococci typically varies from 6 to 64 μg/mL, what sets this antimicrobial ineffective in monotherapy against enterococci. Association of aminoglycoside with an agent active against cell wall formation, such as a β-lactam or glycopeptides, results in a synergistic effect.

However, strains presenting high-level resistance to aminoglycosides, initially described in the 80s, have been observed with increasing frequency around the world. This phenotype has a clinical impact, once high-level resistance to gentamicin is pointed by many studies as an independent predictor of mortality in invasive enterococcal infections. This acquired resistance is generally due to enzymatic inactivation through the production of aminoglycoside-modifying enzymes (AME), which are codified by genes localized in transferable plasmids and transposons, conferring resistance to a few or all representatives of the class. When these enzymes are expressed, aminoglycoside MIC may achieve 2000 µg/mL or higher. Basically, AME may have activity of phosphotranseferases, nucleotidyl transferases or acetyltransferases and it is not uncommon the occurrence of clinical isolates carrying more than one gene, codifying different enzymes.

The *aac(6')-Ie-aph(2'')-Ia* gene codifies a bifunctional enzyme, the AAC(6')-APH(2''), which presents either acetylation and phosphorylation activity. Possibly, this gene came from the fusion of two ancestor genes and it characterizes resistance to a broad spectrum of aminoglycosides, including gentamycin, tobramycin, amikacin, kanamycin, netilmicin and dibekacin. Streptomycin activity is preserved. This enzyme phosphorylates the 2'hydroxy position of gentamicin and simultaneously acetylates the 6'hydroxyl position of other aminoglycosides. Besides enterococci, this gene has already been detected in *S. aureus*, *Staphylococcus epidermidis* and many plasmids of streptococci. The *aac(6')-Ie-aph(2'')-Ia* gene is most commonly flanked by an insertion sequence (IS*256*) in a composite transposon designated Tn*4001* in *S. aureus* and Tn*5281* in *E. faecalis*. More than 90% of enterococcal clinical isolates presenting high-level resistance to aminoglycoside carry this gene.

Several other genes that confer gentamicin resistance have been identified. In comparison to *aac(6')-Ie-aph(2'')-I*, these genes are minor contributors to gentamicin resistance in enterococci. Their prevalence varies by geographical region. The *aph(2'')-Ic* codifies a phosphotransferase related to clinical resistance to gentamicin, tobramycin, kanamycin and dibekacin, but not to amikacin or netilmicin. This gene was initially observed in conjugative plasmids of *E. gallinarum*, and, after that, it has been detected in *E. faecalis* and *E. faecium*, especially in strains from animal sources. In the presence of this gene, gentamicin MIC varies from 256 to 384 µg/mL. Despite the fact that MIC are lower compared to the ones observed to *aac(6')-Ie-aph(2'')-Ia*, *Enterococcus spp.* harboring *aph(2'')-Ic* are clinically resistant to the synergistic effect of gentamicin + ampicillin.

The phosphotransferase codified by *aph(2'')-Id* confers resistance to gentamicin, tobramycin, kanamycin, netilmicin and dibekacin. This gene was initially characterized in an isolate of *E. casseliflavus*, but all further descriptions were in vancomycin-resistant *E. faecium*. Similarly, gene *aph(2'')-Ib* presents activity against the same substrates and is found in the same species. Another gene related to aminoglycoside resistance is *aph(3')-IIIa*, which characterize high-level resistance to kanamycin and avoids synergistic effect of the ampicillin-amikacin association, despite the fact it leads to lower MIC, compared to other AME (64 to 256 µg/mL), which is very similar to the *ant(4')-Ia* activity. This last gene is, by far, less prevalent in clinical isolates and confers resistance to tobramycin, kanamycin and dibekacin. As Aph(3')-IIIa and Ant(4'')-Ia do not confer resistance to gentamicin or streptomycin, they are of less clinical significance.

Unlike other aminoglycoside, high-level resistance to streptomycin may be related to alterations in the ribosomal 30S subunit, leading to reduced binding of the antimicrobial to the target site. *E. faecalis* strains that present this resistance mechanism have streptomycin MIC of 128.000 µg/mL. High-levels resistance to streptomycin also may occur due to the presence of nucleotidyl transferases, codified by *ant(6)-Ia* and *ant(3'')-Ia*. In these cases, streptomycin MIC are lower than previous ones: 4.000 a 16.000 µg/mL.

Significant advances in the knowledge about simultaneous mechanisms of resistance to MLS$_B$ (Macrolides-lincosamides-B streptogramins) group and their genetic determinants were obtained in the last years. The MLS$_B$ group act interfering in different moments of synthesis of bacterial proteins. The main resistance mechanism to macrolides among enterococci is the change of the binding target, which is due to the methylation of an adenine residue of 23S rRNA of the 50S ribosomal subunit. This alteration decreases binding of antimicrobial to ribosome, not only of macrolides, but also of lincosamides and B streptogramins, leading to the MLS$_B$ phenotype. Genetically, there are many *erm* genes responsible for the codification of these methylases, and in *Enterococcus spp.* the mechanism is related, in general, to the presence of *erm*(B) and, rarely, to *erm*(A).

On the other hand, the *mef*(A) gene codifies an efflux protein, active against macrolides, but not lincosamides and B streptogramins, conferring the so-called M phenotype. This gene seems to be located in an conjugative plasmid and mediates lower erythromycin levels (MIC from 2 to 16 µg/mL) if compared to *erm*(B)-related MIC. A third gene, *msr*(A), confers resistance to macrolides and B streptogramins, by encoding an transporter protein related to ATP and it has been detected in clinical isolates of *E. faecium*.

Although macrolides and lincosamides are not commonly used to treat enterococcal infections because of the intrinsic resistance to clindamycin and the high levels of acquired resistance to macrolides, the combination of streptogramins B (quinupristin) and A (dalfopristin) has been an option in cases of multiresistant *E.faecium*. Both components act synergistically, once streptogramin A binds bacteria ribosome leading to conformational alterations that increase affinity of streptogramin B to its target site (50S subunit), inhibiting protein synthesis. Virtually all *E. faecalis* are intrinsically resistant to streptogramin A, which turns ineffective the combination of both, A and B streptogramins. On the other hand, resistance among *E. faecium* varies from 1 to 12% and may be related to the presence of a gene that confers resistance to streptogramin A or to the presence of combined genes, leading to resistance to both molecules. *vat*(D) gene codifies an amino acid sequence closely related to acetyltransferases and mediates resistance to streptogramin A. This gene has been detected in plasmids which also contain *van*A and *erm*(B) genes. A second resistance gene to streptogramin A, *vat*(E), also exhibit an deduced amino acid sequence very similar to acetyltransferases active against streptogramins. Both, *vat*(D) and *vat*(E) genes have been found in *E.faecium* from different sources, including animal and humans. Some other genes [*vgb*(A) and *vat*(A), among other] has also been found in *E. faecium*. However, their frequencies are considerably lower than the above-cited genes. The presence of either *vgb*(A) or *vat*(D) alone may mediates low-level resistance in that full resistance occur when both genes are present.

E.faecium recovered from poultry products have also demonstrated quinupristin/dalfopristin resistance. The use of virginamycin, which is an analogue of A streptogramin used in poultry as a growth promoter in Europe, has probably contributed to the emergence of quinupristin/dalfopristin resistance. Indeed, despite of the fact the prevalence in nosocomial setting is increasing, quinupristin-dalfopristin resistance is most common in environmental samples.

Although not routinely used in treatment of enterococcal infections, tetracyclines present benefic effects in treating bacteremia caused by Vancomycin-Resistant *Enterococcus spp.* (VRE). In these cases, susceptibility tests are indicated to this antimicrobial class. However, in many regions, the majority of clinical isolates present resistance to tetracycline, which do not encourage its use as therapeutical option. There are two major mechanisms of tetracycline resistance: (i) active efflux of the drug throughout the membrane, and (ii) ribosomal protection. The genes *tet*(K) and *tet*(L) codify large proteins within 14 transmembrane domains, leading to resistance due to active expulsion of the antimicrobial. On the other hand, *tet*(M), *tet*(O) and *tet*(S), which are related to tetracycline and minocycline resistance, codify proteins that alter ribosomal conformation, avoiding binding of the antimicrobial to its target site. *tet*(M), the most common tetracycline resistance determinant among enterococci, is typically located on chromosome and usually carried by Tn*916* or related conjugative transposons, but can also be found in conjugative plasmids, justifying the wide dissemination of these gene among enterococci, as well as other bacterial genus.

Likewise tetracyclines, the high prevalence of multiresistant *Enterococcus spp*. in many hospitals have leaded to some interest in the use of chloramphenicol as an alternative therapeutical option. However, despite its infrequent usage, resistance rates to this antimicrobial among enterococci is high. In most cases, resistance is related to specific acetyltransferases, called CAT (Chloramphenicol Acetyltransferases), which acetylate a hydroxyl group on the antimicrobial molecule, avoiding binding to bacterial ribosome. The *cat* genes are usually carried by plasmids, but can also be located on the chromosome.

Quinolone activity against enterococci is considered moderate to high and the resistance to this antimicrobial class has been increasing at the same time than the increase usage of these drugs, especially in genitourinary tract infections. Newer fluoroquinolones, such as moxifloxacin and gatifloxacin present a slightly high activity against enterococci; however, isolates resistant to ciprofloxacin are generally also resistant to moxifloxacin and gatifloxacin. Quinolones act interacting with type II and IV topoisomerases and DNA girase. Both enzymes are essential to the DNA replication. DNA girase (composed by GyrA and GyrB subunits) is the primary target of quinolones in Gram-negative bacteria, while topoisomerase IV (formed by ParC and ParE subunits) is the primary one in Gram-positive bacteria. Mutations in *parC* gene of enterococci may be the first step in the development of quinolone resistance. Further mutations in *gyrA* gene may occur, being associated to higher resistance levels. These mutations may occur in a sequential way as a result of selective pressure. Mutations are almost always related to the Ser83 position of DNA gyrase as well as to the Ser80 position of topoisomerase IV. Mutations in *parC* may lead to low-level resistance, which is increased with the next-step mutation in *gyrA*. Low level resistance may also occur due to alteration in the caption of the antimicrobial into the cell.

In the last two decades, some antimicrobials have demonstrating favorable clinical results for treatment of infections caused by multiresistant *Enterococcus spp.*, especially VRE, to whom therapeutical options are considerably restrict. In this context, linezolid, approved for clinical use in the US in 2000 and for use in United Kingdom a year later, has emerged as a promising option to treat infections caused by multiresistant Gram-positive pathogens, including VRE. Although it is more expensive than vancomycin, linezolid has the advantage of not requiring monitoring serum drug concentrations or dose adjustment for patients with renal or hepatic failure. It is therefore a valuable drug and can be used in situations where vancomycin use is either contraindicated or ineffective. Moreover, it can be given orally due to its high oral bioavailability.

Optimism around linezolid is also because, when the drug was licensed, it represented a new class (oxazolidinones) and, consequently, harbor a new mechanism of action: linezolid is an entirely synthetic drug that binds to the initiation complex and inhibits protein synthesis, being bacteriostatic. Against a new class, bacteria present, in theory, lesser potential to become resistant. However, already in 2001, it has been published the first occurrence of *E. faecium* resistant to linezolid, followed by many other reports, including *E. faecalis* isolates.

In general, this resistance is due to the G2576T mutation in the central loop of domain V of 23S subunit of rRNA. Enterococci produce multiple copies of the gene that codifies the 23S rRNA (*E. faecalis* have four copies and *E. faecium* have six copies). In theory, the presence of multiple gene copies makes resistance from sporadic mutations less likely because the unaffected gene copies would mask the effect of the mutated gene. However, recombination between susceptible and resistant copies (referred as "gene conversion") will yield strains with multiple mutated copies under persistent linezolid selective pressure. Only one mutated allele configure linezolid MIC of 4 to 8 μg/mL; while five mutated alleles increase MIC to 64 μg/mL. Prolonged hospitalization and occurrence and duration of preceding linezolid therapy have been reported as risk factors for the development of resistance, although linezolid-resistant enterococci have also been isolated from patients without any prior therapy with linezolid. Clonal spread of this phenotype has also been described.

Apart of mutations in the 23S rRNA, another mechanism is related to linezolid resistance. A transferable gene, the *cfr*, was identified in *S. aureus* in 2006 as the source of resistance to this antimicrobial, as well as lincosamides and streptogramin A compounds. *cfr* encodes an rRNA methyltransferase that modifies an adenosine in the linezolid-binding region on the 23S rRNA, preventing antibiotic binding. It is hypothesized that the *cfr* gene emerged from animal strains of bacteria that were exposed to natural compounds with an rRNA binding site similar to linezolid. In 2011, *cfr* was identified in an *E. faecalis* strain from a cattle farm in China. In this strain, the gene was located on a plasmid and flanked by IS*1216*. Overall, linezolid resistance remains rare in enterococci.

Daptomycin, a cyclic lipopetide approved by FDA in 2003 for the treatment of complicated skin and soft tissues infections caused by Gram-positive cocci present an excellent activity against enterococci. Indeed, potency of daptomycin and vancomycin are comparable against enterococci. Its bactericidal activity depends on its insertion into the cell membrane in a calcium-dependent manner. It promotes leakage of intracellular potassium into the extracellular space, resulting in cell death by destruction of the transcellular potassium gradient.

Daptomycin has a dose-dependent activity against Gram-positive bacteria. Some large-scale in vitro studies have shown that daptomycin is effective against more than 98% of enterococci tested, irrespective of their susceptibility to other agent. However, failures of daptomycin monotherapy for enterococcal infections have been described in case reports and strains of *Enterococcus spp.* resistant to daptomycin have already been reported in different studies. In general, *E. faecium* is more likely than *E. faecalis* to express daptomycin resistance, although resistance has been reported in both species. The increased prevalence of daptomycin resistance in *E. faecium* may reflect increased use of daptomycin with this species compared with *E. faecalis*, which is usually susceptible to penicillins.

Daptomycin resistance appears to be less common in North America than in Asia or Europe. In *S. aureus*, it have been demonstrated that strains showing reduced susceptibility to daptomycin present mutations in some of the following genes: *mprF, yycG, rpoB* and *rpoC*. These mutations lead to alterations in membrane avoiding activity of the antimicrobial. In enterococci, it seems mutation in some different genes may be involved in resistance. The first gene encodes a putative membrane protein that may be involved in the phosphatidlyserine and sphingolipids biosynthesis, but its function has yet to be determined. The second gene, *cls*, codifies a cardiolipin synthetase, a transphosphatidylase involved in the synthesis of the cell membrane protein, cardiolipin. Also, mutations in *gdpD* (glycerophosphoryl diester phosphodiesterase) and *liaF* (lipid II cycle-interfering diester protein) genes may also be related to resistance to daptomycin. Given that a number of different membrane-associated proteins have been linked to reduced daptomycin susceptibility in staphylococci, it seems likely that more genes conferring enterococcal resistance to daptomycin will be identified in the future.

Especially characterized by its broad spectrum of activity, tigecycline, a glycylglycine derivative of minocycline that prevents elongation of the peptide chain, has shown good activity against *Enterococcus spp.* resistant or susceptible to vancomycin. However, despite the shortened time for clinical use (it was introduced in 2005), resistance among enterococci has already been reported.

Typical tigecycline MIC for enterococci range from 0.125 to 0.25 µg/ml. According to the FDA, MIC ≥ 0.5 µg/ml for *E. faecalis* are considered resistant. The 2013's guideline of Clinical and Laboratory Standards Institute (CLSI) do not contain breakpoints for interpretation of MIC for *Enterococcus spp*. The mechanism of tigecycline resistance in enterococci is unknown. In staphylococci, tigecycline resistance is mediated by a novel family of efflux pumps, but these genes have not been demonstrated in enterococci.

In the context of new antibiotics active against multiresistant *Enterococcus spp*., some new glycopeptides, i.e. telavancin and oritavancin, have shown excellent results against VRE strain, presenting significant pharmacodynamics advantages over vancomycin. Telavancin has been approved for the FDA in 2009 for the treatment of skin and soft tissue infections and in 2013 for the treatment of hospital-acquired and ventilator-associated pneumonia. On the other hand, oritavancin is not approval for clinical use by FDA. Besides, it is necessary to emphasize that, even without approval for clinical use, VRE strains showing reduced susceptibility to oritavancin have been described. Indeed, for the new glycopeptides telavancin, as well as the novel β-lactams ceftobiprole and ceftaroline, only limited clinical data are available.

Ceftaroline and ceftobiprole, fifth generation cephalosporins, have activity against enterococci, but may be prone to emergence of resistance with clinical use. Ceftaroline has been shown to be more efficacious than linezolid in animals and may play a larger role in the future. Ceftobiprole showed high affinity to PBP5 of *E. faecium*, but was withdrawn from the market by the company for further development. Ceftobiprole shows good in vitro activity against *E. faecalis* with no reports of resistance to date, but is ineffective against penicillin-resistant *E. faecium*.

Apart of the new drugs, a number of variable antibiotic combinations have shown in vitro synergistic activity and are promising as potential therapeutical options for VRE infections. Another therapeutical approach for multiresistant enterococci is focused on virulence traits. Some surface molecules (see below) of enterococci have an important role on virulence and strongly contribute to initial adhesion to the tissue or indwelling devices, forming biofilms. Targeting this molecules to inactivate may play a preventive role in enterococcal infections. Further studies must be done to reinforce its application in clinical setting.

Therefore, despite some new therapeutic possibilities directed to multiresistant Gram-positive cocci, treating and controlling infections caused by VRE still remains challenging given the restrict therapeutical options and the disturbing development of resistance to new antimicrobials.

Strains of VRE were first isolated in 1986 in France and England. Since then, enterococci with this phenotype have been described in several parts of the world, presenting different prevalence rates and certain heterogeneity, both phenotypic and genotypic. The most notable consequences of isolation of VRE are increased mortality and prolonged hospital stays, resulting in greater cost. Indeed, resistance to vancomycin has been described as an independent predictor of mortality in invasive enterococcal infections. Furthermore, another worrisome consequence of VRE occurrence is the potential that enterococci have to transfer genetic resistance determinants to other more virulent bacteria, as it was observed within a *S. aureus* strain in 2002 that harbored the *vanA* gene.

The prevalence of VRE has increased dramatically in the 90s. The number of VRE infections in the US hospitals increased from 9,820 in 2000 to 21,352 in 2006. This increase in frequency of VRE isolation have changed the prevalence of species in infections: nowadays, *E. faecium* is now almost as common a cause of nosocomial infections as *E. faecalis*. It is due to the fact that, overall, vancomycin resistance is by far more frequently found among *E. faecium* than *E. faecalis*. If only *E. faecium* is considered, the prevalence of strains resistant to vancomycin in USA is around 50 to 70%. By contrast, only around 5% of *E. faecalis* are vancomycin resistant.

The occurrence of VRE in the US and Europe is consistently diverse over the years. In US VRE has been commonly recovered from inpatients and/or nosocomial environment and less frequently from colonized healthy people, food or animals. Also, the increase in prevalence cited above was firstly observed in US than Europe. Surely, in Europe, the incidence of infections caused by VRE is historically low (around 5%). In European continent, colonization in healthy individuals and its recovery from community sources is much more common than in US, probably due to the use in Europe of avoparcin as a growth promoter in animal husbandry until 1997, when it was banned. Avoparcin is a structural analogue of vancomycin, thus inducing cross-resistance.

However, since the early 1990s and particularly since the beginning of the new century, the prevalence of ampicillin- and vancomycin resistant enterococci has been rising in some European hospitals as a cause of human infections, although important differences do exist in different European regions. By 2007, vancomycin resistance among clinical enterococcal isolates from Europe varied from >30% in countries such as Greece and Ireland to less than 1% in Scandinavian countries, although some recent data from Sweden demonstrated an approximately fourfold increase in infections caused by VRE in 2007-2009 compared to 200-2006. In Asia, VRE occurrence is still low, although some outbreaks have been reported. However, in Latin America, a multicentric study showed a different panorama: despite of the occurrence of VRE, almost 80% of infections are still caused by ampicillin and vancomycin susceptible *E. faecalis*.

Close physical proximity to patients infected or colonized with VRE, prolonged hospitalization, admission to the surgical units or Intensive Care Unit (ICU), solid organ and bone marrow transplantation, co-morbidities such as diabetes, renal failure or hemodialysis, presence of urinary catheter and previous use of antibiotics have previously been associated with VRE acquisition.

However, antimicrobial classes are not equally effective in this selective pressure. Vancomycin and cephalosporins have been pointed as independent risk factors for vancomycin resistance development in many studies, although some authors were not able to assess previous vancomycin use with VRE infection or colonization. On the other hand, treatment with anti- anaerobic drugs have been associated repeatedly with a higher density of colonization with VRE. The anti-anaerobic effect, such as the one observed for metronidazole, suppressed anaerobic bacteria from the gut. In this scenario, enterococci can reproduce easily, as normal microbiota became altered and considerably increase their inoculum. Although somehow controversial, many studies demonstrate a relationship between VRE colonization and further infection with these bacteria.

Vancomycin binds, with high affinity, to the pentapeptide precursor of peptidoglycan. This binding occurs specifically at the C-terminal portion of the dipeptide D-alanine-D-alanine (D-Ala-D-Ala). Thus, transglycosylation blocks the increment of other precursors to nascent peptidoglycan chain, and avoids the subsequent cross-linking of the molecules to form the cell wall by transpeptidation.

Vancomycin resistance is due to the presence of operons, called *van*, encoding enzymes (i) for the synthesis of precursors with low affinity to the antimicrobial, in which the C-terminal dipeptide D-Ala-D-Ala is replaced by other with reduced glycopeptides affinity, such as D-Ala-D-Lac or D-Ala-D-Ser; thereby modifying the binding site of vancomycin, and (ii) for the elimination of high-affinity precursors that are normally produced by the microorganism, removing the binding site of vancomycin. D-Lac residue has been found to have an affinity 1000 times less than D-Ala-D-Ala for vancomycin whereas D-Ala-D-Ser has the affinity about 6 times less than the normal cell wall precursor. For D-Ala-D-Lac, the reduced affinity is because of repulsive forces in the binding pocket of the vancomycin caused by the substitution of the residue D-Ala by D-Lac. On the other hand, for D-Ala-D-Ser, the 6 fold-reduced affinity is associated to the presence of a hydroximethyl group of serine which is bulkier than the methyl group of alanine.

There are nine previously described resistance phenotypes (VanA, VanB, VanC, VanD, VanE, VanG, VanL, VanM and VanN), which present peculiarities considering their genetic determinants and expression of resistance.

VanA type was the first phenotype described. It is associated with high levels of inducible resistance to vancomycin and teicoplanin and configures MIC $\geq$ 64 µg/ml and 16 µg/mL for these antimicrobials, respectively. The first detection of *vanA* gene was in in a plasmid of a clinical isolate of *E. faecium*. VanA resistance is mediated by the transposon *Tn1546* or closely related elements which can be found on plasmids or might be located on chromosomes. This 11kb transposon encodes 9 polypeptides divided into functional groups: transposition (ORF1 and ORF2), regulation of resistance expression (VanR and VanS), synthesis of the dipeptide D-Ala-D-Lac (VanH and VanA), and hydrolysis of the constitutive peptidoglycan precursors (VanX). Another peptide, VanZ, is also encoded by the *vanA* cluster, but its function remains unknown.

During the synthesis of the dipeptide D-Ala-D-Lac, the protein VanH acts as a dehydrogenase, thereby reducing pyruvate to D-Lac while VanA acts as ligase, catalyzing the formation of the bond between D-Ala and D-Lac. The constitutive precursor D-Ala-D-Ala is cleaved by the action of the dipeptidase VanX. The resulting dipeptide replaces the constitutive dipeptide in the synthesis of peptidoglycan, which considerably decreases the affinity of this molecule for glycopeptides.

The *vanA* cluster has been found essentially in *E. faecium* and *E. faecalis*, but also can be observed in *E.avium*, *E.durans*, *E.raffinosus*, and atypical isolates of *E.gallinarum* and

E.casseliflavus, in which high levels of resistance to both vancomycin and teicoplanin can be observed. According to several studies, It is the genetic determinant most commonly associated with vancomycin resistance, regardless the geographical region.

Just as VanA phenotype, VanB acquired resistance is due to the synthesis of peptidoglycan ending in D-Ala-D-Lac. The *vanB* gene is usually located in the bacterial chromosome (conjugative transposons), but can, in specific situations, be found on plasmids. The presence of this gene does not interfere within the susceptibility to teicoplanin but determines inducible resistance to variable, often high, levels of vancomycin, configuring MICs between 4 and 1024 µg/mL. This is because, although the organization and functionality of *vanB* and *vanA* cluster are similar, they differ in their regulation, just as vancomycin and not teicoplanin acts as inducer of *vanB* cluster. Compared to *vanA*, additional open reading frames VanW and VanV have been described on the *vanB* operon; their function remain to be clearly determined.

The operon *vanB* contains genes that codify a dehydrogenase, a ligase and a dipeptidase that present a high degree of sequence homology (67-76% of identity) with the corresponding deduced protein of the *vanA* operon. Based on differences in sequence, the *vanB* gene cluster can be subdivided into three subtypes: *vanB1*, *vanB2* (the most commonly recovered from human sources) and *vanB3*. Nevertheless, no correlation between subtype and level of resistance have been observed. The Tn*1547*, Tn*1549* and Tn*5382* transposon have been described as carriers of the operon *vanB*, and sequencing data suggest that Tn*1549* is essentially identical to Tn*5382*.

Similarly to the above described, the VanD resistance phenotype is due to the production of a precursor of the peptidoglycan ending in D-Ala-D-Lac. The *vanD* gene is located exclusively on chromosome and are not transferable by conjugation which could explain the scarcity of recognized vanD strains in contrast to the widespread and high prevalence of *vanA* and *vanB*. The organization of the operon *vanD*, is similar to the vanA and vanB, presenting some peculiarities. Strains with this phenotype have a small D,D-dipeptidase activity, therefore they are not capable of eliminating the peptidoglycan precursor finalized in D-Ala-D-Ala, target for the action of glycopeptides. Thus, the phenotype VanD is related to moderate levels of resistance to both vancomycin (MIC 64-128 µg/ml) and teicoplanin (MIC 4-64 µg/ml). Based on sequence differences, the *vanD* gene clusters could be separated into subtypes: vanD-1 to vanD-5.

The intrinsic resistance to low levels of vancomycin is a characteristic of *E.gallinarum*, *E. casseliflavus/flavescens*, due to the presence of the *vanC* gene (*E. gallinarum*, vanC1; *E. casseliflavus*, vanC2/vanC3). The deduced protein of VanC2 operon have a high degree of identity (97-100%) with those encoded by the operon VanC3.

The VanC phenotype is expressed constitutively or inducible, producing a precursor of peptidoglycan finalized in D-Ser. The organization of the operon *vanC*, which is chromosomal, differs from the operons described above. Three proteins are required for VanC-type resistance: (i) VanT, a membrane associated serine racemase, which produces D-Ser; (ii) VanC, a ligase which catalyzes the synthesis of D-Ala-D-Ser; and (iii)VanXY$_C$ with dipeptidase and carboxypeptidase activity allowing the hydrolysis of the precursor finished in D-Ala. For many years, it was believed that the operon *vanC* was not transferable. However, recent data have showed the occurrence of *vanC* gene in plasmids of *E. faecalis*. This finding has some important implications: some identification schemes uses the amplification of *vanC* genes for identify *E. gallinarum* and *E. casseliflavus*, which may lead to misidentification.

Like VanC, the phenotype VanE is also related to the synthesis of precursor of peptidoglycan ended in D-Ala-D-Ser and generate low level resistance to vancomycin and susceptibility to teicoplanin. The organization of *vanE* cluster is identical to that observed in VanC operon. It configures an inducible low-level resistance to vancomycin. Another resistance phenotype described in some isolates of *E. faecalis* is VanG. It is an acquired resistance related to the presence of the dipeptide D-Ala-D-Ser that characterizes low levels of resistance to vancomycin (MIC 12 to 16 µg/ml) and teicoplanin susceptibility (MIC lower than 0,5 µg/ml). It is not transferable and it remains to be established if it is a constitutive or inducible phenotype. Based on sequences homologies, *vanG* cluster was subdivided in subtypes: *vanG1* and *vanG2*.

Recently, three new clusters related to vancomycin resistance have been described: *vanL*, *vanM* and *vanN*. The *vanL* gene was characterized in 2008 in a strain of *E. faecalis*, which showed MIC of 8 µg/ml. The expression of resistance is inducible. The *vanL* cluster is essentially organized as *vanC* and *vanE*, and the proteins encoded by their genes catalyze the formation of the dipeptide D-Ala-D-Ser which is incorporated into the peptidoglycan precursor. Since the source of *vanL* gene as well as the ways of acquisition are unknown, the clinical significance of these findings remains to be established.

Six clinical isolates of *E.faecium*, recovered from 2005 to 2008 in China, harbored a new vancomycin resistance related gene, the *vanM*. All strains presented high level resistance to vancomycin (> 256µg/mL) and variable teicoplanin MICs (from 0.75 to higher than 256 µg/mL). This transferable cluster is related to the production of D-Ala-D-Lac dipeptide and it is unclear if the expression of resistance is inducible or constitutive.

The most recently described *van* cluster, *vanN*, was characterized in *E. faecium* strains recovered from domestic chicken meat in Japan. All isolates showed low-level resistance to vancomycin (MIC = 12 µg/ml), susceptibility to teicoplanin and were clonal. The vancomycin resistance among these strains was encoded in a large plasmid and was expressed constitutively. It seems not to be a transferable characteristic and the peptidoglycan precursor produced by this cluster is D-Ala-D-Ser.

Apart of enterococci presenting resistance to vancomycin, It have been described the occurrence of enterococcal strains that require vancomycin to growth. Strains with this phenotype were recovered from patients that had received vancomycin therapy for long periods and most of them harbored the *vanB* gene, although some *vanA* strains have also been described. The vancomycin dependence is related to the fact that constitutive pathway for peptidoglycan formation ending in D-Ala-D-Ala is interrupted. It occurs because the ligase loose its activity and the production of the peptidoglycan precursor ending in D-Ala-D-Ala is interrupted. Therefore, cells have to produce peptidoglycan based on alternative precursors (D-Ala-D-Lac) which does not depend of the activity of the constitutive ligase previously described. Because induction of production of D-Ala-D-Lac precursors only occur in presence of vancomycin, these strains do not growth in the absence of this antimicrobial.

Despite the somewhat well-defined phenotypes of vancomycin resistance, the in vitro characterization of them is not always as simple as it seems. Recent studies have reported strains of VRE isolated in Korea, Japan and Taiwan showing incongruence between phenotype and genotype, which were susceptible to teicoplanin in in vitro tests, despite the presence of the vanA gene. It has been suggested that point mutations may be the cause of the divergence. Similarly, it has been observed (non-published data) some strains of *E. faecalis* harboring and expressing the *vanC* gene. In this scenario, for epidemiological studies, both phenotypic and genotypic features must be considered.

In conclusion, the sum of all these above-cited intrinsic and acquired resistance features frequently leads to multiresistant enterococcal strains, which rapidly disseminates in community and/or hospital environment. Surveillance and epidemiological studies are strongly encouraged with the objective to monitor those strains.

In terms of epidemiology, a few years ago, the enterococcal infections were traditionally considered endogenous, originating from the patient's own normal microbiota. For this reason, the epidemiology of these infections has not arouse much attention. In recent decades, however, great interest was given to this subject due to evidences that have supported the exogenous acquisition of enterococcal infections.

An understanding of the transmission dynamics of enterococci in the hospital environment is pivotal for infection control. Hospitalized patients often receive antimicrobials that increase the density of VRE in the gastrointestinal tract (such as metronidazole mentioned above), in turn facilitating the spread of this organism. An interesting mathematical model of transmission drew parallels between VRE transmission and the transmission of a vector-borne infection, such as malaria. In this model, health-care personnel play the role of the mosquito, carrying VRE on their hands from patients who are VRE positive (infection or colonization) to those who are negative, and to their surroundings, and there is a potential to transmit the pathogens with each contact. Transmission can be amplified depending on how many patients have contact with the contaminated-health-care staff and correlates with the density of VRE in patient stools.

The management of dissemination of VRE requires strategies to contain cases and reduce transmission rates. This concept highlights the importance of curtailing the chain of transmission through (i) active surveillance (surveillance cultures) and contact precautions for infected and colonized individuals, (ii) implementation of strict hand hygiene practices for health-care workers and educational campaigns to support this behavior, (iii) judicious use of antimicrobials and (iv) aggressive environmental cleaning method.

Guidelines for the prevention of nosocomial transmission of VRE have been addressed in several publications and adherence to this guidelines has been shown to decrease nosocomial spread but has failed to eradicate VRE, especially in an endemic setting. On the other hand, some studies suggest that robust infection control strategies enable the timely termination of VRE outbreaks, even those involving strains with high epidemic potential on "high-risk wards", such as hematology and oncology. Premature discontinuation of measures may represent possible causes for recurrence of VRE spread. Besides, failures to completely control VRE may also be explained by prolonged environmental contamination, which is in some situation the source of the dissemination. Large time periods between surveillance cultures, thus allowing colonized patients and environment to be left out of precautions may also represent a cause for VRE recurrence.

As mentioned above, surveillance cultures to identify colonized patients represent one important steps trying to eradicate VRE from the nosocomial environment, once it enable to establish contact precaution measures for patients who carry VRE on their guts, as they are reservoirs for VRE along with infected patients. CDC and the Society of Healthcare Epidemiology of America recognize the importance of active VRE surveillance to reduce or eliminate hospital-acquired infections.

Culture of rectal swabs using selective, enriched or differential media is the method of choice for VRE screening. Currently, one of the most widely used for screening medium is the azide bile-esculin agar supplemented with 6 µg of vancomycin per milliliter. Chromogenic media have been also used in clinical settings.

However, time required for culture results is a subject of major concern. Indeed, any established surveillance program will benefit from the fast identification of VRE carriers by allowing the rapid isolation of those patients, thus minimizing the spread. Rapid identification become even more important in developing countries where isolation rooms are scarce.

The time to get results for culture-based methods varies from 48 to 72h. Besides, because of the high density of normal microbiota in rectal swabs, sensitivity to recover VRE may be low. Therefore, the availability of an assay capable of detecting VRE with good sensitivity in few hours coupled to an appropriate infection control program could reduce hospital-acquired VRE infections, thus leading to a significant reduction of patient mortality by this particular agent.

Several nucleic acid amplification tests have been developed and evaluated for the detection of VRE. The most commonly used are the ones based on conventional or real-time PCR reactions. Limitations of these tests include the following. (i) the requirement of complex extraction and detection procedures or the requirement of a culture step when testing is done from a selective enrichment broth or isolates are recovered from solid media; (ii) high cost per sample and expensive equipment; and (iii) the most widely used molecular approach (e.g. detection of *vanA-vanB* by real time PCR) deliver less information than the culture test (e.g. species identification and detection of *vanC* genotypes), which limits the full integration of molecular approaches for VRE detection. Irrespective the methodology chosen by microbiology laboratory, infection control staff must be awarded about limitations of it to better interpret results.

Multiresistant-enterococcal outbreaks can be either related to the dissemination of a single strain, following a clonal pattern, and be associated to the horizontal transference of genetic elements carrying resistance genes, generating a heterogeneous scenario with the participation of multiple clones.

The development of molecular typing systems contributed significantly to the understanding and tracking the spread of multiresistant-enterococcal strains. Such systems have varying degrees of discriminatory power and reproducibility. Therefore different molecular typing methods have present distinct purposes and the knowledge of the technique will enable more certainty when choosing the typing method.

Over the years, several of these methods have been developed. Some examples include: analysis of the electrophoretic polymorphism of isoenzymes (MLEE), ribotyping, and DNA amplification techniques based on the methodology of the polymerase chain reaction with its different variants. However, these techniques have present reduced discriminatory powder and/or reproducibility.

Currently, there are basically two gold-standard techniques for enterococcal molecular typing: Pulsed-Field Gel Electrophoresis (PFGE) and Multilocus Sequence Typing (MLST). PFGE is based on comparison of band pattern generated by digestion of the whole bacterial chromosome by using a restriction enzyme with rare targets across the bacterial chromosome. Band patterns can be visualized after an electrophoresis in a pulsed field, as large fragments would not be effectively separated using a conventional electrophoresis methodology.

PFGE is very discriminatory but present some limitations on reproducibility, once electrophoretic conditions are somewhat difficult to standardize between different laboratories. For regional purposes of outbreak characterization, PFGE have excellent utility. However, there is no a worldwide database to globally understand dissemination of some specific clones.

On the other hand, MLST generates unambiguous data, easily comparable from any geographical region. This typing technique is based on sequencing of internal fragments of seven housekeeping genes (*gdh, gyd, pst, gki, aroE, xpt, yiq* for *E. faecalis*). Each allele of each gene receive a number and the combination of the seven allele number generates a sequence type (ST), which are grouped in clonal complexes (CC). A database is available (www.mlst.net), where it is possible to obtain the worldwide behavior of a certain strain.

By using these methodologies, many data has been available about the way enterococci, and especially multiresistant strains, disseminates. The clonal transmission was confirmed, for example, to the majority of β-lactamase-producers enterococcal outbreaks. Dissemination of high-level gentamicin resistant enterococci present both clonal and heterogeneous pattern. When non-clonal strains are part of an outbreak, endogenous acquisition is more probable.

VRE outbreaks may present polyclonal origin, although its dissemination is often clonal. This is because, once established within a given hospital, these clones become extremely difficult to eradicate.

Early studies on the molecular epidemiology of *E. faecalis* and *E. faecium* using PFGE showed the spread of single clones to different US states. MLST results demonstrated that enterococcal isolates recovered from hospitalized patients often cluster in specific groups. Indeed the majority of hospital-related *E. faecalis* grouped in two clonal complexes: CC2 and CC9. On the other hand, the worldwide increase prevalence of *E. faecium* in hospitals is mainly due to CC17, especially ST17, ST18, ST78 and ST192. Recent comparisons of available genome sequences support the concept of a hospital-associated clade, which is genetically distinct from most commensal isolates from animals and humans.

One of the main challenges in the analysis of enterococci is the enormous plasticity of their genome, where acquired elements can account for up to 25% of the genome in both *E. faecalis* and *E. faecium*. Another important component in the evolution of multidrug-resistant strains of hospital-associated enterococci may be their lack of CRISPR (clustered regulatory interspaced short palindromic repeats) elements, which provide bacteria with a defense system against incoming DNA. Indeed, the presence of antibiotic resistance determinants is inversely correlated with the presence of CRISPRs in *E. faecalis*, and most isolates from hospital-associated clonal complexes lack CRISPRs.

In conclusion, either because of their physiological characteristics and their epidemiological behavior, *E. faecalis* are a subject of major concern particularly in hospital environments. Knowledge about mechanism of resistance to most recent drugs along with the understanding of epidemiological dynamics of certain clones (and the recognition of specific characteristics of those clones, such as virulence and resistance traits) can strongly support infection control team on the implementation of more effective measures to avoid enterococcal dissemination.

REFERENCES

Aarestrup, F.M.; Butaye, P. & Witte, W. 2002. Nonhuman reservoirs of enterococci. p 55-99. *In*: Gilmore, M.S., Clewell, D.B., Courvalin, P., Dunny, G.M., Murray, B.E. & Rice, L.B. *The enterococci: pathogenesis, molecular biology, and antibiotic resistance*. ASM Press, Washington DC. P. 55-100.

Abbassi, M.S.; Achour, W. & Hassen, A.B. 2007. High-Level gentamicin-resistant *Enterococcus faecium* strains isolated from bone marrow transplant patients: accumulation of antibiotic resistance genes, large plasmids and clonal strain dissemination. *Int. J. Antimicrob. Agents*. 29: 658-664.

Agerso, Y.; Pedersen, A.G. & Aarestrup, F.M. 2006. Identification of Tn5397-like and Tn916-like transposons and diversity of the tetracycline resistance gene tet (M) in enterococci from humans, pigs and poultry. *J. Antimicrob. Chemother*. 57: 832-839.

Alburquerque, V.S.; Silva, C.M.F.; Marques, E.A.; Teixeira, L.M. & Merquior, V.L.C. 2000. Occurrence of vancomycin-resistant *Enterococcus faecalis* in Rio de Janeiro, Brazil: strains showed genetic relationship with a high-level gentamicin resistant (HLGR) endemic clone. p. 107. Abstract of 40th Interscience Conference on Antimicrobial Agents and Chemotherapy.

Anderegg, T.R.; Sader, H.S.; Fritsche, T.R.; Ross, J.E. & Jones, R.N. 2005. Trends in linezolid susceptibility patterns: report from the 2002-2003 worldwide Zyvox Annual Appraisal of Potency and Spectrum (ZAAPS) Program. *Int. J. Antimicrob. Agents*. 26: 13-21.

Anderson, D.J.; Murdoch, D.R.; Sexton, D.J.; Reller, L.B.; Stout, J.E.; Cabell, C.H. & Corey, G.R. 2004. Risk factors for infective endocarditis in patients with enterococcal bacteremia: a case-control study. *Infection* 32: 72-77.

Anderson, D.J.; Olaison, J.R.; Miro, J.M.; Hoen, B.; Selton-Suty, C.; Doco-Lecompte, T.; Abrutyn, E.; Habibi, G.; Eykyn, S.; Pappas, P.A.; Fowler, V.G.; Sexton, D.J.; Almela, M., Corey, G.R. & Cabell, C.H. 2005. Enterococcal prosthetic valve infective endocarditis: report of 45 episodes from International Collaboration on Endocarditis-merged database. *Eur. J. Clin. Microbiol. Infect*. Dis. 24: 665-670.

Andrade, S.S.; Sader, H.S.; Jones, R.N.; Pereira, A.S.; Pignatari, A.C.C. & Gales, A.C. 2006. Increased resistance to first-line agents among bacterial pathogens isolated from urinary tract infections in Latin America: time for local guidelines? *Mem. Inst. Oswaldo Cruz* 101: 741-748.

Antalek, M.D.; Mylotte, J.M.; Lesse, A.J. & Sellick JR., J.A. 1995. Clinical and molecular epidemiology of *Enterococcus faecalis* bacteremia, with special reference to strains with high-level resistance to gentamicin. *Clin. Infect. Dis*. 20: 103-109.

Arias, C.A., Murray, B.E. 2012. Attributable costs of enterococcal bloodstream infections in a nonsurgical hospital cohor. *Nat. Rev. Microbiol*. 16: 266-278.

Byers, K.E.; Anglim, A.M.; Anneski, C.J. & Farr, B.M. 2002. Duration of colonization with vancomycin-resistant enterococcus. *Infect. Control Epidemiol*. 23: 207-211.

Boyd, D.A.; Willey, B.M.; Fawcett, D.; Gillani, N. & Mulvey, M.R. 2008. Molecular characterization of Enterococcus faecalis N06-0364 with low-level vancomycin resistance harboring a novel D-Ala-D-Ser gene cluster, vanL. *Antimicrob. Agents Chemother*. 52: 2667-72.

Butt, T., Leghari, M.J. & Mahmood, A. 2004. In-vitro activity of nitrofurantoin in enterococcus urinary tract infection. *J. Pak. Med. Assoc.* 54: 466-469.

Busani, L.; Del Grosso, M.; Paladini, C.; Graziani, C.; Pantosti, A.; Biavasco, F. & Caprioli, A. 2004. Antimicrobial susceptibility of vancomycin-susceptible and –resistant enterococci isolated in Italy from raw meat products, farm animals, and human infections. *Int. J. Food Microbiology* 97: 17-22.

Caiaffa Filho, H.H.; Almeida, G.D.; Oliveira, G.A.; Sarahyba, L.; Mamizuka, E.M. & Burattini, M.N. 2003. Molecular characterization of *van* genes found in vancomycin-resistant Enterococcus spp. isolated from the Hospital das Clínicas, FMUSP, São Paulo, Brazil. *Braz. J. Infect. Dis.* 7: 173-174.

Camargo, I.I.B.C.; Barth, A.L.; Pilger, K.; Seligman, B.G.S.; Machado, A.R.L. & Darini, A.L.C. 2004. *Enterococcus gallinarum* carrying the *vanA* gene cluster: first report in Brazil. *Braz. J. Med. Biol. Res.* 37: 1669-1671.

Camargo, I.L.; Gilmore, M.S. & Darini, A.L. 2006. Multilocus sequence typing and analysis of putative virulence factors in vancomycin-resistant and vancomycin-sensitive *Enterococcus faecium* isolates from Brazil. *Clin. Microbiol. Infect.* 12: 1123-1130.

Canalejo, E.; Ballesteros, R.; Cabezudo, J.; García-Arata, M.I. & Moreno, J. 2008. Bacteraemic spondylodiscitis caused by *Enterococcus hirae*. *Eur. J. Clin. Microbiol. Infect.* Dis. 27: 613-615.

Cano, M.E.; Domínguez, M.A.; Ezpeleta, C.; Padilla, B.; De Arellano, E.R. & Martinez-Martinez, L. 2008. Cultivos de vigilancia epidemiológica de bactérias resistentes a los antimicrobianos de interés nosocomial. *Enfer. Infecc. Microbiol. Clin.* 26: 220-229.

Carvalho, M.G.S. 1998. Caracterização bio-epidemiológica de *Enterococcus* e microrganismos relacionados, com ênfase na aplicação de métodos moleculares. Tese de Doutorado. IM, UFRJ, RJ, Brasil.

Carvalho, M.G.S.; Teixeira, L.M. & Facklam, R.R. 1998. Use of tests for acidification of methyl-α-D-glucopyranoside and susceptibility to efrotomycin for differentation of strains of *Enterococcus* and some related genera. *J. Clin. Microbiol.* 36: 1584-1587.

Carvalho, M.G.S.; Steigerwalt, A.G.; Morey, R.E.; Shewmaker, P.L.; Teixeira, L.M. & Facklam, R.R. 2004. Characterization of Three new enterococcal species, *Enterococcus spp.* nov. CDC PNS-E1, *Enterococcus spp*.nov. CDC PNS-E2, *Enterococcus spp.* nov. CDC PNS-E3, isolated from human clinical specimens. *J. Cli. Microbiol.* 42: 1192-1198.

Carvalho, M.G.; Steigerwalt, A.G.; Morey, R.E.; Shewmaker, P.L.; Falsen, E.; Facklam, R.R. & Teixeira, L.M. 2008. Designation of the provisional new Enterococcus sppecies CDC PNS-E2 as Enterococcus sanguinicola sp. nov., isolated from human blood, and identification of a strain previously named Enterococcus CDC PNS-E1 as Enterococcus italicus Fortina, Ricci, Mora and Manachini 2004. *J. Clin. Microbiol.* 46: 3473-3476.

Casadewall, B. & Courvalin, P. 1999. Characterization of the *vanD* glycopeptide resistance gene cluster from *Enterococcus faecium* BM4339. *J. Bacteriol.*, 181: 3644-3648.

Cassone, M.; Del Grosso, M.; Pantosti, A.; Giordano, A. & Pozzi, G. 2008. Detection of genetic elements carrying glycopeptide resistance clusters in *Enterococcus* by DNA microarrays. *Mol. Cel.Probes* 22: 162-167.

Cetinkaya, Y.; Falk, P. & Mayhall, C.G. 2000. Vancomycin-resistant enterococci. *Clin. Microbiol. Rev.* 13: 686-707.

Cercenado. E. 2011. *Enterococcus*: phenotype and genotype resistance and epidemiology in Spain. *Enferm. Infecc. Microbiol.* Clin. 29: 59-65.

Cereda, R.; Pignatari, A.C.; Hashimoto, A. & Sader, H. 1997. *In vitro* antimicrobial activity against enterococci isolated in an university hospital in São Paulo, Brazil. *Braz. J. Infect. Dis.* 1: 83-90.

Cereda, R.F.; Gales, A.C.; Silbert, S.; Jones, R.N. & Sader, H.S. 2002. Molecular typing and antimicrobial susceptibility of vancomycin-resistant *Enterococcus faecium* in Brazil. *Infect. Control Hosp. Epidemiol.* 23: 19-22.

Chow, J.W.; Thal, L.A.; Perri, M.B.; Vazquez, J.A.; Donabedian, S.M.; Clewell, D.B. & Zervos, M.U. 1993. Plasmid-associated hemolysin and aggregation substance production contribute to virulence in experimental enterococcal endocarditis. *Antimicrob. Agents Chemother.* 37: 2474-2477.

Christiansen, K.J.; Turnidge, J.D.; Bell, J.M.; George, N.M.; Pearson, J.C. & Australian Group On Antimicrobial Resistance. 2007. Prevalence of antimicrobial resistance in *Enterococcus* isolates in Autralia, 2005: report from the Australian Group on Antimicrobial Resistance. *Commun. Dis. Intell.* 31: 392-397.

Clark, N.C.; Cooksey, R.C.; Hill. B.C.; Swenson, J.M. & Tenover, F.C. 1993. Characterization of glycopeptide-resistant enterococci from U.S. hospitals. *Antimicrob. Agents Chemother.* 37: 2311-2317.

Clark, N.C.; Teixeira, L.M.; Facklam, R.R. & TENOVER, F.C. 1998. Detection and differentiation of *vanC1, vanC-2* and *vanC-3* glycopeptide resistance genes in enterococci. *J. Clin. Microbiol.* 36: 2294-2297.

Clinical And Laboratory Standards Institute. 2008. Performance standards for antimicrobial susceptibility testing, seventeenth informational supplement, document M100-S17. CLSI, Wayne, Pa, USA.

Cobo Molinos, A.; Abriouel, H.; Omar, N.B.; López, R.L.; Galvez, A. 2008. Detection of ebp (endocarditis- and biofiol- associated pilus) genes in enterococcal isolates from clinical and non-clinical origin. *Int. J. Food Microbiol.* 126: 123-126.

Colurn, P.S.; Pillar, C.M.; Jett, B.D.; Haas, W. & Gilmore, M.S. 2004. *Enterococcus faecalis* senses target cells and in response expresses cytolysin. *Science* 306: 2270-2272.

Comert, F.B.; Kulah, C.; Aktas, E.; Ozlu, N. & Celebi, G. 2007. First isolation of vancomycin-resistant enterococci and spread of a single clone ina a university in northwestern Turkey. *Eur. J. Clin. Microb. Infect. Dis.* 26: 57-61.

Coque, T.M.; Patterson, J.E.; Steckelberg, J.M. & Murray, B.E. 1995. Incidence of hemolysin, gelatinase, and aggregation substance among enterocicci isolated from patients with endocarditis and other infections and from feces of hospitalized patients and community-based persons. *J. Infect. Dis.* 171: 1223-1229.

Coque, T.M.; Tomayko, J.F.; Ricke, S.C.; Okhyusen, P.C. & Murray, B.E. 1996. Vancomycin-resistant enterococci from nosocomial, community, and animal sources in the United States. *Antimicrob. Agents Chemother.* 40: 2605-2609.

Corso, A.C.; Gagetti, P.S.; Rodríguez, M.M.; Melano, R.G.; Ceriana, P.G.; Faccone, D.F.; Galas, M.F. & Vre Argentinean Collaborative Group. 2007. Molecular epidemiology of vancomycin resistant *Enterococcus faecium* in Argentina. *Int. J. Infect. Dis.* 11: 69-75.

Courvalin, P. 2006. Vancomycin resistance in Gram-positive cocci. *Clin. Infect. Dis.* 42:S25-S34.

Couto, R.C.; Carvalho, E.A.A.; Pedrosa, T.M.G.; Pedroso, E.R.; Neto, M.C. & Biscione, F.M. 2007. A 10-year prospective surveillance of nosocomial infections in neonatal intensive care units. *Am. J. Infect. Control* 35: 183-189.

Cuzon, G.; Naas, T.; Fortineau, N. & Nordmann, P. 2008. A novel Chromogenic medium for detection of Vancomycin-resistant *Enterococcus faecium* e *E. faecalis*. *J. clin. Microbiol.* 46: 2442-2444.

Dahl, K.H.; Simonsen, G.S.; Olsvik, O. & Sundsfjord, A. 1999. Heterogeneity in the *vanB* gene cluster of genomically diverse clinical strains of vancomycin-resistant enterococci. *Antimicrob. Agents Chemother.* 43: 1105–1110.

D'agatha, E.M.; Gerrits, M.M.; Tang, Y.W.; Samore, M. & Kusters, J.G. 2001. Comparison of pulsed-field gel electrophoresis and amplified fragment-length polymorphism for epidemiological investigations of common nosocomial pathogens. *Infect. Control Hosp. Epidemiol.* 22: 550-554.

D'agatha, E.M.C.; Gautam, S.; Green, W.K. & Tang, Y.W. 2002. High rate of false-negative results of the rectal swabs culture method in detection of gastrintestinal colonization with vancomycin-resistant enterococci. *Clin. Infect. Dis.* 34: 167-172.

Daikos, G.L.; Bamias, G.; Kattamis, C.; Zervos, M.J.; Christakis, G.; Petrikkos, G.; Triantafyllopoulou, P.; Alexandrou, H. & Syriopoulou, V. 2003. Structure, locations, and transfer frequencies of genetic ements conferring hogh-level gentamicina resistance in *Enterococcus faecalis* insolates in Greece. *Antimicrob. Agents Chemother.* 47: 3950-3953.

Dalla Costa, L.M.; Souza, D.C.; Martins, L.T.F.; Zanella, R.C.; Brandileone, M.C.; Bokermann, S.; Sader, H.S. & Souza, A.P.H.M. 1998. Vancomycin-resistant *Enterococcus faecium*: first case in Brazil. *Braz. J. Infect. Dis.* 2: 160-163.

Darini, A.L.; Palepou, M.F. & Woodford, N. 2000. Effects of the movement of insertion sequences on the structure of VanA glycopeptide resistance elements in *Enterococcus faecium*. *Antimicrob. Agents Chemother.* 44: 1362-1364.

Da Silva, P.S.; Monteiro Neto, H. & SEJAS, L.M. 2007. Successful treatment of vancomycin-resistant enterococcus ventriculitus in a child. *Braz. J. Infect. Dis.* 11: 297-299.

d'azevedo, P.A.; Kacman, S.B.; Schmalfuss, T. & Rodríguez LM. 2000. Primeiro caso de *Enterococcus* resistente a vancomicina isolado em Porto Alegre, RS. *J. Brás. Patol. Méd. Labo.* 36: 258.

d'azevedo, P.A.; Cantarelli, V.; Inanime, E.; Superti, S.; Dias, C.A.G. 2004. Avaliação de um sistema automatizado na identificação de espécies de *Enterococcus*. *J. Bras. Pat. Med. Lab.* 48: 217-219.

D'azevedo PA, DIAS CAG & TEIXEIRA LM. 2006 Genetic diversity and antimicrobial resistance of Enterococcal isolates from Southern region of Brazil. *Rev. Inst. Méd. Trop.* 48: 11-16.

d'azevedo, P.A.; Furtado, G.H.; Medeiros, E.A.; Santiago, K.A.; Silbert, S. & Pignatari, A.C. 2008. Molecular characterization of vancomycin-resistant enterococci strains eight years apart from its first isolation in São Paulo, Brazil. *Rev. Inst. Med. Trop. São Paulo*, 50: 195-198.

De Leener, F.; Martel, A.; Decostere, A.; Haesebrouck, F. 2004. Distribution of the *erm*(B) gene, tetracycline resistance genes and *Tn1545*-like transposons in macrolide- and lincosamide resistant enterococci from pigs and humans. *Microb. Drug Resist.* 10: 341-345.

De Niederhäusern, S.; Sabia, C.; Messi, P.; Guerrieri, E.; Manicardi, G. & Bondi, M. 2007. vANa-TYPE Vancomycin-resistant enterococci in equine and swine rectal swabs and in human clinical samples. *Curr. Microbiol.* 55: 240-246.

Depardieu, F.; Perichon, B. & Courvalin, P. 2004. Detection of the van alphabet and identification of enterococci and staphylococci at the species level by multiplex PCR. *J. Clin. Microbiol.* 42: 5857-5860.

Dervisoglou, A.; Tsiodras, S.; Kanellakopoulou, K., Pinis, S.; Galanakis, N.; Pierakakis, S.; Giannakakis, P; Liveranou, S.Ntasiou, P.; Karampali, E.; Iordanou, C. & Giamarellou, H. 2006. The value of chemoprophylaxis against *Enterococcus spp*ecies in elective cholecystectomy. *Arch. Surg.* 141: 1162-1167.

Deshpande, L.M.; Fritsche, T.R.; Moet, G.J. Biedenbach, D.J. & Jones, R.N. 2007. Antimicrobial resistance and molecular epidemiology of vancomycin-resistant enterococci from North America and Europe: a report from the SENTRY antimicrobial surveillance program. *Diagn. Microb. Infect. Dis.* 58: 163-170.

Devriese, L.A.; Vancanneyt, M.; Descheemaeker, P.; Baele, M., Van Landuyt, H.W.; Gordts, B.; Butaye, P.; Swings, J. & Haesebrouck, F. 2002. Differentiation and identification of *Enterococcus durans*, *E. hirae* and *E. villorum*. *J. Applied Microbiology* 92: 821-827.

Diazgranado, C.A.; Zimmer, S.M.; Klein, M. & Jernigan, J.A. 2005. Comparison of mortality associated with vancomycin-resistant and vancomycin-susceptible enterococcal bloodstream infections: a meta-analysis. *Clin. Infect. Dis.* 41: 327-332.

Domig, K.J. Mayer, H.K. & Kneifel, W. 2003. Methods used for the isolation, enumeration, characterization and identification of Enterococcus spp. 2. Pheno- and genotpypic criteria. *Int. J. Food Microbiol.* 88: 165-188.

Donabedian, S.M.; Thal, L.A.; Hershberger, E.; Perri, M.B.; Chow, J.W.; Bartlett, P. Jones, R.; Joyce, K.. Rossiter, S.; Gay, K.; Johnson, J.; Mackinson, C.; Debess, E.; Madden, J.; Angulo, F. & Zervos, M.J. 2003. Molecular characterization of gentamicina-resistant *Enterococci* in the United States: evidence of spread from animas to humans through food. J. Clin. Microbiol. 41: 1109-1113. *Arch. Surg.* 141: 1162-1167.

Draghi, D.C.; Benton, B.M.; Krause, K.M.; Thornsberry, C.; Pillar, C. & Sahm, D.F. 2008. In vitro activity of telavancin against recent Gram-positive clinical isolates: results of the 2004-05 Prospective European Surveillance Initiative. *J. Antimicrob. Chemother.* 62: 116-121.

Duh, R.W.; Singh, K.V.; Malathum, K. & Murray, B.E. 2001. *In vitro* activity of 19 antimicrobial agents against enterococci from healthy subjects and hospitalized patients and use of an *ace* gene probe from *Enterococcus faecalis* form species identification. *Microbiol. Drug Resist.* 7: 39-46.

Dunny, G.M.; Leonard, B.A. & Hedberg, P.J. 1995. Pheromone-inducible conjugation in *Enterococcus faecalis*: interbacterial and host-parasite chemical communication. *J. Bacteriol.* 177: 871-876.

Dutta, I. & Reynolds, P.E. 2003. The vanC-3 vancomycin resistance gene cluster of *Enterococcus flavescens* CCM439. *J. Antimicrob. Chemother.* 51: 703-706.

Eaton, T.J. & Gasson, M.J. 2001. Molecular screening of *Entercoccus* virulence determinants and potential for genetic exchanges between food and medical isolates. *Appl. Envirn. Microbiol.* 67: 1628-1635.

Edmond, M.B.; Ober, J.F.; Weinbaum, D.L.; Pfaller, M.A.; Hwang, T.; Sanford, M.D. & Wenzel, R.P. 1995. Vancomycin-resistant *Enterococcus faecium* bacteremia: risk factors for infection. *Clin. Infect. Dis.*, 20: 1126-1133.

Eliopoulos, G.M. 1993. Aminoglycoside-resistant enterococcal endocarditis. *Infect. Dis. Clin. North. Am.* 7: 117-133.

Emaneini, M.; Aligholi, M. & Aminshahi, M. 2008. Characterization of glycopeptides, aminoglycosides and macrolide resistance among *Enterococcus faecalis* and *Enterococcus faecium* isolates from hospitals in Tehran. *Pol. J. Microbiol.* 57: 173-178.

Endtz, H.P.; Van Den Braak, N.; Van Belkun, A.; Kluytmans, J.A.; Koeleman, J.G.; Spanjaard, L.; Voss, A.; Weersink, A.J.; Vandenbroucke-Grauls, C.M.; Buiting, A.G.; Van Duin, A. & Verbrugh, H.A. 1997. Fecal carriage of vancomycin-resistant enterococci in hospitalized patients and those living in the community in the Netherlands. *J. Clin. Microbiol.* 35: 3026-3031.

Endtz, H.P.; Van Den Braak, N.; Van Belkum, A.; Geessens, W.H. Kreft, D.; Streebel, A.B. & Verbrugh, A. 1998. Comparison of eight methods to detect vancomycin resistance in enterococci. *J. Clin. Microbiol.* 36: 592-594.

Evers, S. & Courvalin, P. 1996. Regulation of the VanB-type resistance gene expression by the $VanS_B$ -$VanR_B$ two-component regulatory system in *Enterococcus faecalis* V583. *J. Bacteriol.* 178: 1302-1309.

Facklam, R.R.& Collins, M.D. 1989. Identification of *Enterococcus spp*ecies isolated from human infections by a conventional test scheme. *J. Clin. Microbiol.* 27: 731-734.

Facklam, R.R. & Teixeira, L.M. 1998. Enterococcus. In: Collier, L., Ballows, A. & Susman, M. (ed). *Topley and Wilson's Microbiology and Microbial Infections*. 9 th ed. Edward Arnold. London, United Kindon. p. 669-682.

Facklam, R.R.; Sahm, D. F. & Teixeira, L. M. 1999. *Enterococcus*. In: Murray P.R, Baron E.J., Pfaller M.A., Tenover F.C., Yolken R.H. (eds). *Manual of Clinical Microbiology.* 7th ed. ASM Press, Washington, DC. p.297-305.

Facklam, R.R.; Carvalho, M.G.S. & Teixeira, L.M. 2002. History, taxonomy, biochemical characteristics, and antibiotic susceptibility testing of enterococci.. In: Gilmore, M.S. Clewell, D.B., Courvalin, P., Dunny, G.M., Murray, B.E.& Rice, L.B. *The enterococci: pathogenesis, molecular biology, and antibiotic resistance*. ASM Press, Washington D/C. p. 1-54.

Fadda, G.; Nicoletti, G.; Schito, G.C.; Tempera, G. 2005. Antimicrobial susceptibility patterns of contemporary pathogens from uncomplicated urinary tract infections isolated in a multicenter Italian survey: possible impact on guidelines. *J. Chemother.* 17: 251-257.

Fines, M.; Perichon, B.; Reynolds, P.; Sahm, D.F. & Courvalin, P. 1999. VanE, a new type of acquired glycopeptide resistance in *Enterococcus faecalis* BM4405. *Antimicrob Agents Chemother.* 43: 2161-2164.

Falk, P.S.; Winnike, J.; Woodmansee, C.; Desani, M. & Mayhall, C.G. 2000. Outbreak of vancomycin-resistant enterococci in a burn unit. *Infect. Control Hosp. Epidemiol.* 21: 575-582.

Fontana, R.; Ligozzi, M.; Pittaluga, F. & SATTA, G. 1996. Intrinsic penicillin resistance in enterococci. *Microb. Drug Resist.* 2: 209-213.

Fortina, M.G.; Ricci, G.; Mora, D. & Manachini, P.L. 2004. Molecular analysis of artisanal Italian cheeses reveals Enterococcus italicus sp. nov. *Int. J. Syst. Evol. Microbiol.* 54: 1717-1721.

Fortún, J.; Coque, T.M.; Martín-Dávila, P.; Moreno, L.; Cantón, R.; Loza, E.; Baquero, F. & MORENO, S. 2002. Risk factors associated with ampicillin resistance inpatients with bacteremia caused by *Enterococcus faecium. J. Antimicrob. Chemother.* 50: 1003-1009.

Freeman, C.; Robinson, A.; Cooper, B.; Mazens-Sullivan, M.; Quintiliani, R. & Nightingale, C. 1993. In vitro antimicrobial susceptibility of glycopeptide-resistant enterococci. *Diag. Microbiol. Infect.* Dis. 21: 47-50.

Freitas, M.C.; Pacheco-Silva, A.; Barbosa, D.; Silbert, S.; Sader, H.; Sesso, R. & Camargo, L.F. 2006. Prevalence of vancomycin-resistant enterococcus fecal colonization among kidney transplant patients. *BMC Infect. Dis.* 22: 133.

Furtado, G.H.C.; Martins, S.T.; Coutinho, A.P.; Soares, G.M.M.S.; Wey, S.B. & Servolo, E.A. 2005a. Incidence of vancomycin-resistant *Enterococcus* at a university hospital in Brazil. Rev. Saúde Pública 39: 1-5.

Furtado, G.H.; Martins, S.T.; Coutinho, A.P.; Wey, S.B. & Medeiros, E.A. 2005b. Prevalence and factors associated with rectal vancomycin-resistant enterococci colonization in two intensive care units in São Paulo, Brazil. *Braz. Infect. Dis.* 9: 64-69.

Furtado, G.H.; Mendes, R.F.; Pignatari, A.C.; Wey, S.B. & Medeiros, E.A. 2006. Risk factors for vancomycin-resistant *Enterococcus faecalis* bacteraemia in hospitalized patients: an analysis of two case-control studies. *Am. J. Infect. Control.* 34: 447-451.

Garcia-Migura, L.; Liebana, E. & Jensen, L.B. 2007. Transposon characterization of vancomycin-resistant *Enterococcus faecium* (VREF) and dissemination of resistance associated with transferable plasmids. *J. Antimicrob. Chemother.* 60: 263-268.

Gikas, A.; Christidou, A.; Scoulica, E.; Nikolaidis, P.; Skoutelis, A.; Levidiotou, S.; Kartali, S.; Maltezos, E.; Metalidis, S.; Kioumis, J.; Haliotis, G.; Dia, S.; Roumbelaki, M.; Papageorgiou, N.; Kritsotakis, E.I. & Tselentis, Y. 2005 Epidemiology and molecular analysis of intestina colonization by vancomycin-resistant enterococci in greek hospitals. *J. Clin. Microbiol.* 43: 5796-5799.

Gilad, J.; Borer, A.; Riesenberg, K.; Peled, N.; Shnaider, A. & Schlaeffer, F. 1998. *Enterococcus hirae* septicemia in a patient with end-stage renal disease undergoing hemodialysis. *Eur. J. Clin. Microbiol. Dis.* 17: 576-577.

Gilmore, M.S.; Segarra, R.A.; Booth, M.C.; Bogie, C.P.; Hall, L.R. & Clewell, D.B. 1994. Genetic structure of the *Enterococcus faecalis* plasmid pAD1-encoded cytilytic toxin system and its relationship to lantibiotic determinants. *J.Bacteriol.* 176: 7335-7344.

Gilmore, M.S.; Coburn, P.S.; Nallapareddy, S.R. & Murray, B.E. 2002. Enterococcal virulence.. In: Gilmore, M.S. Clewell, D.B., Courvalin, P., Dunny, G.M., Murray, B.E.& Rice, L.B. *The enterococci: pathogenesis, molecular biology, and antibiotic resistance.* ASM Press, Washington DC. p 301-54.

Giraffa, G. 2002. Enterococci from foods. FEMS Microbiol. Rev. 26: 163-171.

Gonzales, R.D.; Schreckenberger, P.C.; Graham, M.B.; Kelkar, S.; Denbesten, K. & Quinn, J.P. 2001. Infections due to vancomycin-resistant *Enterococcus faecium* resistant to linezolid. *Lancet* 14: 1179.

Gordillo, M.E.; Singh, K.V.; Baker, C.J. & Murray, B.E. 1993. Typing of Group B streptococci: comparison of pulsed-field gel electrophoresis and conventional electrophoresis. *J. Clin. Microbiol.* 31: 1430-1434.

Gordillo, M.E.; Singh, K.V. & Murray, B.E.1993. Comparison of ribotyping and pulsed-field gel electrophoresis for subspecies differentiation of strains of *Enterococcus faecalis. J. Clin. Microbiol.* 31: 1570-1574.

Gordon, S.; Swenson, J.M.; Hill, B.C.; Pigott, N.E.; Facklam, R.R.; Cooksey, R.C.; Thornsberry, C.; Enterococcal Study Group; Jarvis, W.R. & Tenover, F.C. 1992. Antimicrobial susceptibility patterns of common and unusual species of enterococci causing infections in the United States. *J. Clin. Microbiol.* 30: 2373-2378.

Green, M.R. Anasetti, C., Sandin, R.L., Rolfe, N.E. & Greene, J.N. 2006. Development of daptomycin resistance in a bone marrow transplant patient with vancomycin-resistant Enterococcus durans. *J. Oncol. Pharm.* Pract. 12: 179-181.

Grinhaum, R.S.; Gimarães, T.; Kusano, F.; Hosino, N.; Sader, H. & Cereda, R.F. 2003. A pseudo-outbreak of vancomycin-resistant *Enterococcus faecium*. *Infect. Control. Hosp. Epidemiol.* 24: 461-464.

Guerrero, F.; Manuel, L.; Goyenechea, A.; Verdejo, C.; Roblas, R.F. & Gorgolas, M. 2007. Enterococcal Endocarditis on Native and Prosthetic Valves: A Review of Clinical and Prognostic Factors With Emphasis on Hospital-Acquired Infections as a Major Determinant of Outcome. *Medicine* 86: 363-377.

Guven, M.; Bulut, Y.; Sezer, T.; Aladag, I.; Eyibilen, A. & Etikan, I. 2006. Bacterial etiology of acute otitis media and clinical efficacy of amoxicillin-clavulanate versus azithromycin. *Int. J. Pediatr.* Otorhinolaryngol 70: 915-923.

Haas, W.; Shepard, B.D. & Gilmore, M.S. 2002. Two-component regulator of *Enterococcus faecalis* cytolysin responds to quorum–sensing autoinduction. *Nature* 415: 84-87.

Hall, L.M.C.; Duke, B.; Urwin, G. & Guiney, M. 1992. Epidemiology of *Enterococcus faecalis* urinary tract infection in a teaching hospital in London, United Kingdom. *J. Clin. Microbiol.* 30: 1953-1957.

Hancock, L.E., & Gilmore, M.S. 2000. Pathogenicity of enterococci. *In*: Fischetti, V.A. & Novick, R.P. (ed.) *Gram-positive pathogens.*: ASM Press. Washington, DC. P. 251-258.

Handwerger, S.; Raucher, B.; Altarac, D.; Monka, J.; Marchione, S.; Singh, K.V.; Murray, B.E.; Wolff, J. & Walters, B. 1993. Nosocomial outbreak due to *Enterococcus faecium* highly resistant to vancomycin, penicillin and gentamicin. *Clin. Infect. Dis.* 16: 750-755.

Hannecart-Pokorni, E.; Depuydt, F.; De Wit, L.; Van Bossuyt, E.; Content, J. & Vanhoof, R. 1997. Characterization of 6'N-aminoglycoside acetyltransferase gene *aac(6')I* associated with a *sull*-type integron. *Antimicrob. Agents Chemother.* 41: 314-318.

Higashide, T.; Takahashi, M.; Kobayashi, A.; Ohkubo, S.; Sakurai, M.; Shirao, Y.; Tamura, T. & Sogiyama. 2005. Endophthalmitis caused by *enterococcus mundtii*. *J. Clin. Microbiol.* 43: 1475-1476.

Hill, E.E.; Herijgers, P.; Claus, P.; Vanderschueren, S.; Herregods, M.C. & Peetermans, W.E. 2007. Infective endocarditis: changing epidemiology and predictors of 6-month mortality: a prospective cohort study. *Eur. Hearth Journal* 28: 196-203.

Hollenbeck, B.L.; Rice, L.B. 2012. Intrinsic and acquired resistance mechanisms in *Enterococcus*. *Virulence* 15: 421-433.

Hörner, R.; Liscano, M.G.H.; Maraschin, M.M.; Salla, A.; Meneghetti, B.; Forno, N.F.D. & Righi, R.A. 2005. Susceptibilidade antimicrobiana entre amostras de *Enterococcus* isoladas no Hospital Universitário de Santa Maria. *Jornal Brasileiro de Patologia e Medicina Laboratorial* 41: 391-395.

Hsueh, P.R., Teng, L.J., Chen, Y.C., Yang, P.C., Ho, S.W. & Luh, K.T. 2000. Recurrent bacteremic peritonitis caused by *Enterococus cecorum* in a patient with liver cirrhosis. *J. Clin. Microbiol.* 38: 2450-2452.

Huycke, M.M., Spiegel, C.A., & Gilmore, M.S. 1991. Bacteremia caused by hemolytic, high-level gentamicin-resistant *Enterococcus faecalis*. *Antimicrob. Agents Chemother.* 35:1626-1634.

Huycke, M.H., Sahm, D.F. & Gilmore, M.S. 1998. Multiple-drug resistant enterococci: the nature of the problem and an agenda for the future. *Emerg. Infect. Dis.* 4: 239-249.

Iaria, C.; Stassi, G.; Costa, G.B., Di Leo, R.; Toscano, A.; CASCIO.A. 2005. Enterococcal meningitis caused by *Enterococcus casseliflavus*. First case report. *BMC Infect. Dis.* 14: 3.

Ike, Y. & Clewell, D.B. 1992. Evidence that the hemolysin/bacteriocin phenotype of *Enterococcus faecalis* subsp. *Zymogenes* can be determined by plasmids in different incompatibility groups as well as by the chromosome. *J. Bacteriol.* 174: 8172-8174.

Iwen, P.; Rupp, M.E.; Schreckenberger, P.C. & Hinrichs, S.H. 1999. Evaluation of the revised MicroScan Dried Overnight Gram-positive identification panel to identify *Enterococcus spp*ecies. *J. Clin. Microbiol.* 37: 3756-3758.

Jayaratne, P. & Rutherford, C. 1999. Detection of clinically relevant genotypes of vancomycin-resistant enterococci in nosocomial surveillance specimens by PCR. *J. Clin. Microbiol.* 37: 2090-2092.

Jesudason, Mv; Pratima, Vl; Pandian, R, Abigail, S. 1998. Characterization of penicillin resistant enterococci. *Indian J. Med. Microbiol.* 16: 16-18.

Jett, B.D., Huycke, M.M., & Gilmore, M.S. 1994. Virulence of enterococci. *Clin Microbiol. Rev.* 7: 462-478.

Jones, R.N.; Beach, M.L.; Pfaller, M.A. & Doern, G.V. 1998. Antimicrobial activity of gatifloxacin tested against 1676 strains of ciprofloxacin-resistant gram-positive cocci isolated from patient infection in North and South America. *Diagn. Microbiol. Infect. Dis.* 32: 247-252.

Jones, M.E.; Gesu, G.; Ortisi, G.; Sahm, D.F.; Critchky, I.A. & Goglio, A. For The Assiciazione Microbiologi Clinici Italiani (Amcli) Committee For Antibiotics. 2002. Proficiency of italian clinical laboratories in detectiong reduced glycopeptide susceptibility in *Enterococcus* and *Staphylococcus* spp. using routine laboratory methodologies. *Clin. Microbiol. Infect.* 8: 101-111.

Jones, S.; England, R.; Eyans, M.; Soo, S.S.; Venjatesan, P. 2007. Microbiologically confirmed meningoencephalitis due to *Enterococcus avium*: a first report. *J. Infect.* 54: 129-131.

Kainer, M.A.; Devasia, R.A.; Jones, T.F. Simmons, B.P.; Melton, K. Chow, S.; Broyles, J.; Moore, K.L.; Craig, A.S. & Schaffner, W. 2007. Response to emerging infection leading to outbreak of linelozid-resistant enterococci. *Emerg. Infect. Dis.* 13: 1024-1030.

Kak, V. & Chow, J.W. 2002. Acquired antibiotic resistance in enterococci. *In*: Gilmore, M.S. Clewell, D.B., Courvalin, P., Dunny, G.M., Murray, B.E.& Rice, L.B. *The enterococci: pathogenesis, molecular biology, and antibiotic resistance.* ASM Press, Washington DC. p 355-83.

Kanematsu, E.; Deguchi, T.; Yasuda, M.; Kawamura, T.; Nishino, Y. & Kawada, Y. 1998. Alterations in the GyrA subunit of DNA gyrase and the ParC subunit of DNA topoisomerase IV associated with quinolone resistance in *Enterococcus faecalis*. *Antimicrob. Agents Chemother.* 42: 433-435.

Kaplan, A.H.; Gilligan, P.H. & Facklam, R.R. 1988. Recovery of resistant enterococci during vancomycin prophylaxis. *J. Clin. Microbiol.* 26: 1216-1218.

Kariyama, R.; Mitsuhata, R.; Chow, J.W.; Clewell, D.B. & Kumon, H. 2000. Simple and reliable multiplex PCR assay for surveillance of isolates of vancomycin-resistant enterococci. *J. Clin. Microbiol.* 38: 3092-3095.

Klibi, N.; Gharbi, S.; Masmoudi, A.; Bem Slama, K.; Poeta, P.; Zarazaga, M.; Fendri, C.; Boudabous, A. & Torres, C. 2006. Antibiotic resistance and mechanisms implicated in clinical enterococci in a Tunisian hospital. *J. Chemother.* 18: 20-26.

Klingenberg, R.; Schmied, C.; Falk, V.; Lüscher, T.F.; Zbinden, A.; Corti, R. 2012. Molecular evidence of *Enterococcus faecalis* in an occluding coronary thrombus from a patient with late prosthetic valve endocarditis. *Clin. Res. Cardiol.* 101: 1021-1023.

Ko, Ming-Chung, Liu, Chih-Kuang, Woung, Lin-Chung, Lee, Wen-Kai, Jeng, Huey-Sheng, Lu, Shing-Hwa, Chiang, Hian-Sun & Li, Chung-Yi. 2008. Species and Antimicrobial Resistance of Uropathogens Isolated from Patients with Urinary Catheter. Tohoku *J. Exp. Med.* 214: 311-319.

Koch, S.; Hufnagel, M.; Theilacker, C. & Huebner, J. 2004. Enterococcal infections: host response, therapeutic and prophylatic possibilities. *Vaccine* 22: 822-830.

Kolar, M.; Pantucek, R.; Bardon, J.; Cekanova, L.; Kesselova, M.; Sauer, P.; Vagberova, I. & Koukalova, D. 2005. Occurrence of vancomycin-resistant enterococci in human and animals in the Czech Republic between 2002 and 2004. *J. Med. Microbiol.* 54: 965-967.

Koort, J.; Coenye, T.; Vandamme, P.; Sukura, A. & Bjorkroth, J. 2004. Enterococcus hermanniensis sp. nov., from modified-atmosphere-packaged broiler meat and canine tonsils." *Int. J. Syst. Evol. Microbiol.* 54:1823-1827.

Kuriyama, T.; Willams, D.W.; Patel, M.; Lewis, M.A.O.; Jenkins, L.E.; Hill, D.W. & Hosein, I.K. 2003. Molecular characterization of clinical and environmental isolates of vancomycin-resistant *Enterococcus faecium* e *Enterococcus faecalis* from a teaching hospital in Wales. *J. Med. Microbiol.* 52: 821-827.

Kurup, A.; Chlebicki, M.P.; Ling, M.L.; Koh, T.H.; Tan, K.Y.; Lee, L.C. & Howe, K.B.M. 2008. Control of a hospital-wide vancomycin-resistant *Enterococci* outbreak. *Am. J. Infect. Control.* 36: 206-211.

Lambiase, A.; Del Pezzo, M.; Piazza, O.; Petagna, C.; De Luca, C. & Rossano, F. 2007. Typing of vancomycin-resistant *Enterococcus faecium* strains in a cohort of patients in an italian intensive care unit. *Infection* 35: 428-433.

Laupland, K.B.; ROSS, T.; PITOUT, J.D.; CHURCH, D.L. & GREGSON, D.B. 2007. Community-onset urinary tract infections: a population-based assessment. *Infection* 35: 150-153.

Law-Brown, J. & Meyers, P.R. 2003. Enterococcus phoeniculicola sp. nov., a novel member of the enterococci isolated from the uropygial gland of the Red-billed Woodhoopoe, Phoeniculus purpureus. *Int. J. Syst. Evol. Microbiol.* 53:683-685.

Leavis, H.L.; Bonten, M.J.M. & Willems, R.J.L. 2006. Identification of high-risk enterococcal clonal complexes: global dispersion and antibiotic resistance. *Curr. Opin. Microbiol.* 9: 454-460.

Leclercq, R.; Derlot, E.; Duval, J. & Courvalin, P. 1988. Plasmid-mediated resistance to vancomycin and teicoplanin in *Enterococcus faecium*. *N. Engl. J. Med.* 319: 157-161.

Leclercq, R.; Dutka-Malen, S.; Duval, J. & Courvalin, P. 1992. Vancomycin resistance gene *vanC* is specific to *Enterococcus gallinarum*. *Antimicrob. Agents Chemother.* 36: 2005-2008.

Lee, W.G.; Huh, J.Y.; Cho, S.R. & Lin, Y.A. 2004. Reduction in glycopeptide resistance in vancomycin-resistant enterococci as a result of vanA cluster rearrangements. *Antimicrob. Agents Chemother.* 48: 1379–81.

Lentino, J.R.; Narita, M. & Yu, V.L. 2008. New antimicrobial agents as therapy for resistant gram-positive cocci. *Eur. J. Clin. Microbiol. Infect.* Dis. 27: 3-15.

Lester, C.H.; Fridmodt-Moller, N. & Hammerum, A.M. 2004. Conjugal transfer of aminoglycoside and macrolide resistance between *Enterococcus faecium* isolates in the intestine of streptomycin-trated mice. *FEMS Microbiol Lett.* 235: 385-391.

Lewis, J.S.; Owens, A.; Cadena, J.; Sabol, K.; Patterson, J.E. & Jorgensen, J.H. 2005. Emergence of daptomycin resistance in *Enterococcus faecium* during daptomycin therapy. *Antimicrob. Agents Chemother.* 49: 1664-1665.

Ligozzi, M.; Pittaluga, F. & Fontana, R. 1996. Modification of penicillin-binding protein 5 associated with high-level ampicillin resistance in *Enterococcus faecium. Antimicrob. Agents Chemother.* 40: 354-357.

López, M.; Cercenado, E.; Tenorio, C.; Ruiz-Larrea, F.; Torres, C. 2012. Diversity of clones and genotýpes among vancomycin-resistant clinical *Enterococcus* isolates recovered in a Spain hospital. *Microb. Drug Resist.* 18: 484-491.

López, F.; Culebras, E.; Betriú, C.; Rodriguez-Avial, I.; Gómez, M. & Picazo, J.J. 2010. Antimicrobial susceptibility and macrolide resistance genes in Enterococcus faecium with reduced susceptibility to quinupristin-dalfopristin: level of quinupristin-dalfopristin resistance is not dependent on erm(B) attenuator region sequence. *Diagn. Microbiol. Infect. Dis.* 66: 73-77.

Luna, V.A.; Coates, P.; Eady, E.A.; Cove, J.H.; Nguyen, T.T.H. & Roberts, M.C. 1999. A variety of gram-positive bacteria carry mobile *mef* genes. *J. Antimicrob. Chemother.* 44: 19-25.

Lyerová, L.; Viklický, O.; Nemcová, D. & Teplan, V. 2008. The incidence of infectious diseases after renal transplantation: a single-centre experience. *Int. J. Antimicrob. Agents* 31: S58-62.

Ma, X.; Michiaki, K.; Takahashi, A.; Tanimoto, K. & IKE, Y. 1998. Evidence of nosocomial infection in Japan caused by high-level gentamicin-resistant *Enterococcus faecalis* and identification of the pheromone-responsive conjugative plasmid encoding gentamicin resistance. *J. Clin. Microbiol.* 36: 2460-2464.

Macovei, L. & Zurek, L. 2006. Ecology of antibiotic resistance genes: characterization of enterococci from houseflies collected in food settings. *Appl. Environ. Microbiol.* 72: 4028-4035.

Magliano, E.; Grazioli, V.; Deflorio, L.; Leuci, A.I.; Mattina, R.; Romano, P.; Cocuzza, C.E. 2012. Gender and age-dependent etiology of commynity-acquired urinary tract infections. *Scientific World Journal* 12: 349597.

Malani, P.N., Kauffman, C.A. & Zervos, M.J. 2002. Enterococcal disease, epidemiology and treatment. *In*: Gilmore, M.S. Clewell, D.B., Courvalin, P., Dunny, G.M., Murray, B.E.& Rice, L.B. *The enterococci: pathogenesis, molecular biology, and antibiotic resistance.* ASM Press, Washington DC. p 385-408.

Malathum, K.; Singh, K.V.; Weinstock, G.M. & Murray, B.E. 1998. Repetitive sequence-based PCR versus pulsed-field gel electrophoresis for typing of *Enterococcus faecalis* at the subspecies level. *J. Clin. Microbiol.* 36: 211-215.

Malthotra-Kumar, S.; Haccuria, K.; Michiels, M.; Ieven, M.; Poyart, C.; Hryniewicz, W. & Goossens, H. 2008. Current trends in Rapid Diagnostics for Methicillin-Resistant *Staphylococcus aureus* and Glycopeptide-Resistant *Enterococcus*. *J.Clin. Microb.* 46: 1577-1587.

Manson, J.A.; Keis, S.; Smith, J.M. & Cook, G.M. 2003. A clonal lineage of VanA-type *Enterococcus faecalis* predominates in vancomycin-resistant enterococci isolated in New Zeland. *Antimicrob. Agents and Chemother.* 47: 204-210.

Markowitz, S.M.; Wells, V.D.; Williams, D.S.; Stuart, C.G.; Coudron, P.E. & Wong, E.S. 1991. Antimicrobial susceptibility and molecular epidemiology of β-lactamase-producing, aminoglycoside-resistant isolates of *Enterococcus faecalis*. *Antimicrob. Agents Chemother.* 35: 1075-1080.

Martinez-Odrizola, P.; Muñoz-Sánchez, J.; Gutiérrez-Macias, A.; Arriola-Martinez, P.; Montero-Aparicio, E.; Ezpeleta-Baquedano, C.; Cisterna-Cáncer, R. Miguel De La Villa, F. 2007. Na analysis of 182 enterococcal bloodstream infections: epidemiology, microbiology, and outcome. *Enferm. Infecc. Microbiol.* Cli. 25: 503-507.

Maschieto, A.; Martinez, R.; Palazzo, I.C.V.; Darini, A.L.C. 2004. Antimicrobial Resistance to *Enterococcus spp.* isolated from the intestinal tract of patients from a university hospital in Brazil. Mem. *Inst. Oswaldo Cruz* 99: 763-767.

Mascini, E.M.; Bonten, M.J. 2005. Vancomycin-resistant enterococci: consequences for therapy and infection control. *Clin. Microbiol. Infect.* 11 (supl. 4): 43-46.

Mckessar, S.J.; BERRY, A.M.; BELL, J.M.; TURNIDGE, J.D. & PATON, J.C. 2000. Genetic characterization of *vanG*, a novel resistance lócus of *Enterococcus faecalis*. *Antimicrob. Agents Chemother.* 44: 3224-3228.

Megran, D.W. 1992. Enterococcal Endocarditis. Clin. Infect. Dis. 80: 63-71.

Merquior, V.L.C.; Peralta, J.M.; Facklam, R.R. & Teixeira, L.M. 1994. Analysis of electrophoretic whole-cell protein profiles as a tool for characterization of *Enterococcus sppecies*. *Curr. Microbiol.* 28: 149-153.

Merquior, V.L.; Netz, D.J.; Camello, T.C. & Teixeira, L.M. 1997. Characterization of enterococci isolated from nosocomial and community infections in Brazil. *Adv. Exp. Med. Biol.* 418: 281-283.

Merquior, V.L.; Gonçalves Neves, F.P.; Ribeiro, R.L.; Duarte, R.S.; De Andrade Marques, E. & Teixeira, L.M. 2008. Bacteraemia associated with a vancomycin-resistant *Enterococcus gallinarum* strain harbouring both the *vanA* and *vanC1* genes. *J. Med. Microbiol* 57: 244-245.

Młynarczyk, G.; Grzybowska, W.; Młynarczyk, A.; Tyski, S.; Kawecki, D.; Łuczak, M.; Chmura, A. & Rowiński, W. 2007. Significant increase in the isolation of glycopeptide-resistant enterococci from patients hospitalized in the transplant surgery ward in 2004-2005. *Transplant Proc.* 39: 2883-5.

Mondino, S.S.B.; Castro, A.C.D.; Mondino, P.J.; Carvalho, M.G.; Silva, K.M. & Teixeira, L.M. 2003. Phenotypic and genotypic characterization of clinical and intestinal enterococci isolated from inpatients and outpatients in two brazilian hospitals. *Microbiol. Drug. Resist.* 9: 167-74.

Morrison, D. Woodford, W. & Cookson, B. 1997. Enterococci as emerging pathogens of humans. *J. Appl. Microbiol.* 83: S89-SS99.

Mundy, L.M.; Sahm, D.F., & Gilmore, M.S. 2000. Relationships between enterococcal virulence and antimicrobial resistance. *Clin. Microbiol. Rev.* 13: 513-522.

Munita, J.M.; Arias, C.A.; Murray, B.E. 2012. Enterococcal endocarditis: can we win the war? *Curr. Infect. Dis. Rep.* 14: 339-349.

Munoz-Price, L.S.; Lolans, K. & Quinn, J.P. 2005. Emergence of resistance to daptomycin during treatment of vancomycin-resistant *Enterococcus faecalis* infection. *Clin. Infect. Dis.* 41: 565-566.

Murao, S.; Hosokawa, H.; Hosokawa, Y.; Ishida, T. & Takahara, J. 1997. Discitis, Infectious arthritis, and bacterial meningitis in a patient with pancreatic diabetes. *Internal Medicine* 36: 443-445.

Murray, B.E. 1990. The life and times of the *Enterococcus. Clin. Microbiol. Rev.* 3: 46-65.

Murray, B.E. 1991. New aspects of antimicrobial resistance and the resulting therapeutic dilemmas. *J. Infect. Dis.* 163: 1185-1194.

Murray, B.E.; Lopardo, H.A.; Rubeglio, E.A.; Frosolono, M. & Singh, K.V. 1992. Intra-hospital spread of a single gentamicin-resistant, β-lactamase-producing strain of *Enterococcus faecalis* in Argentina. *Antimicrob. Agents Chemother.* 36: 230-232.

Murray, B.E. 2000. Vancomycin-resistant enterococcal infections. *N. Eng. J. Med.* 342: 710-721.

Naber, K.G.; Schito, G.; Botto, H.; Palou, J.; Mazzzei, T. 2008. Surveillance study in Europe and Brazil on clinical aspects and antimicrobial resistance epidemiology in female with cystitis (ARESC): implications for empiric therapy. *Eur. Urol.* 54: 1164-1175.

Nallapareddy, S.R.; Duh, R.W.; Singh, K.V.; Murray Be. 2002. Molecular typing of selected Enterococcus faecalis isolates: pilot study using multilocus sequence typing and pulsed-field gel electrophoresis. *J. Clin. Microbiol.* 40: 868-876.

Ng, L.-K.; Martin, I.; Alfa, M. & Mulvey, M. 2001. Multiplex PCR for the detection of tetracycline resistant genes. *Mol. Cell. Probes* 15: 209-215.

Nishimoto, Y.; Kobayashi, N.; Alam, M.M.; Ishino, M.; Uehara, N. & Watanabe, N. 2005. Analysis of the prevalence of tetracycline resistance genes in clinical isolates of *Enterococcus faecalis* and *Enterococcus faecium* in a Japanese hospital. *Microb. Drug Resist.* 11: 146-153.

Novais, C.; Freitas, A.R.; Souza, J.C.; Baquero, F.; Coque, T.M. & Peixe, L.V. 2008. Diversity of Tn*1546* and its role in the dissemination of vancomycin-resistant enterococci in Portugal. *Antimicrob. Agents Chemother.* 52: 1001-1008.

Nomura, T.; Tanimoto, K.; Shibayama, K.; Arakawa, Y.; Fulimoto, S.; Ike, Y.; Tomita, H. 2012. Identification of vanN-type vancomycin resistance in an *Enterococcus faecium* isolate from chicken meat in Japan. *Antimicrob. Agents and Chemother.* 56: 6389-6392.

Ntokou, E.; Stathopoulos, C.; Kristo, I.; Dimitoulia, E.; Labrou, M.; Vasdeki, A.; Makris, D.; Zakynthinos, E.; Tsakris, A.; Pournaras, S. 2012. Intensive care unit dissemination of multiple clones of linezolid-resistant *Enterococcus faecalis* and *Enterococcus faecium. J. Antimicrob. Chemother.* 67: 1819-1823.

Oh, W.S.; Ko, K.S.; Song, J.H.; Lee, M.Y.; Park, S.; Peck, K.R.; Lee, N.Y.; Kim, C.; Lee, H.; Kim, S.; Chang, H.; Kim, Y.; Jung, S.; Son, J.S.; Yeom, J.; Ki, H.K. & Woo, G. 2005. High rate of resistance to quinupristin–dalfopristin in *Enterococcus faecium* clinical isolates from Korea. *Antimicrob. Agents Chemother.* 49: 5176–5178.

Ortu, M.; Gabrielli, E.; Caramma, I.; Rossotti, R.; Gambirasio, M. & Gervasoni, C. 2008. *Enterococcus gallinarum* endocarditis in a diabetic patient. *Diabetes Res. Clin, Pract.* 81: e18-21.

Papaparaskevas, J.; Vatopoulos, A.; Tassios, P.T.; Avlami, A.; Legakis, N.J. & Kalapothaki, V. 2000. Diversity among high-level aminoglycoside-resistant enterococci. *J. Antimicrob. Chemother.* 45: 277-83.

Patel, R.; Uhl, J.R.; Hopkins, M.K. & Cockerill, F.R. 1997. Multiplex PCR detection of *vanA*, *vanB*, *vanC*-1, and *vanC*-2/3 genes in enterococci. *J. Clin. Microbiol.* 35:703-707.

Paula, G. R., 2000. Análise da diversidade genética de amostras de enterococos obtidas no Hospital Universitário Clementino Fraga Filho da Universidade Federal do Rio de Janeiro, de 1995 a 1997. Tese de Mestrado. UFRJ, RJ, Brasil.

Pelicioli Riboldi, G.; Preusser De Mattos, E.; Guedes Frazzon, A.P. Alves D'azevedo, P.A. & Frazzon, J. 2008. Phenotypic and genotypic heterogeneity of *Enterococcus sp*ecies isolated from food in Southern Brazil. *J. Basic Microbiol.* 48: 31-37.

Pereira, G.H.; Müller, P.R.; Zanella, R.C., De Jesus Castro Lima, M.; Torchio, D.S.; Levin, A.S. 2010. Outbreak of vancomycin-resistant enterococci in a tertiary hospital: the lack of effect of measures directed mainly by surveillance cultures and differences in response between *Enterococcus faecium* and *Enterococcus faecalis*. *Am. J. Infect. Control.* 38: 406-409.

Poeta, P.; Costa, D.; Igrejas, G.; Rojo-Bezares, B.; Sáenz, Y.; Zarazaga, M.; Ruiz-Larrea, F.; Rodrigues, J. & Torres, C. 2007. Characterization of vanA-containing *Enterococcus faecium* isolates carrying Tn*5397*-like and Tn*916*/Tn*1545*-like transposons in wild boars. *Microb. Drug Resist.* 13: 151-156.

Pourshafie, M.R.; Talebi, M.; Saifi, M. Katouli, M. ; Eshraghi, S.; Kühn, I. & Möllby, R. 2008. Clonal heterogeneity of clinical isolates of vancomycin-resistant *Enterococcus faecium* with unique vanS. *Trop. Med. Int. Health.* 13: 722-727.

Portillo, A.; Ruiz-Larrea, F.; Zarazaga, M.; Alonso, A.; Martinez, J.L. & Torres, C. 2000. Macrolide resistanse genes in Enterococcus spp. *Antimicrob. Agents Chemother.*,44: 967-971.

Prakash, V.P.; Rao, S.R.; Parija, S.C. 2005. Emergence of unusual species of enterococci causing infections, South India. *BMC Infect. Dis.* 5: 5-14.

Qu, T.T. ; Chen, Y.G.; Yu, Y.S.; Wei, Z.Q.; Zhou, Z.H. & Li, L.J. 2006. Genotypic diversity and epidemiology of high-level gentamicin resistant *Enterococcus* in a Chinese hospital. *J. Infect.* 52: 124-130.

Quiñones D, Goñi P, Rubio Mc, Duran E, Gómez-Lus R. 2005. Enterococci spp. isolated from Cuba: species frequency of occurrence and antimicrobial susceptibility profile. *Diag. Microbiol. Infect. Dis.* 51: 63-67.

Quintiliani, R.; Evers, S. Jr. & Courvalin, P. 1993. The *vanB* gene confers various levels of self-transferable resistance to vancomycin in enterococci. *J. Infect. Dis.* 167: 1220-1223.

Raveh, D.; Rosenzweig, I.; Rudensky, B.; Wiener-Well, Y. & Yinnon, A.M. 2006. Risk factors for bacteriuria due to *Pseudomonas aeruginosa* and *Enterococcus spp* in patients hospitalized via the emergency departament. *Eur. J. Clin. Microb. Infect. Dis.* 25: 331-334.

Reis, A.O.; Cordeiro, J.C.; Machado, A.M. & Sader, H.S. 2001. In vitro antimicrobial activity of linezolid tested against vancomycin-resistant enterococci isolated in brazilian hospitals. *Braz. J. Infect. Dis.* 5: 243-251.

Reik, R.; Tenover, F.C.; Klei, E. & Mcdonald, L.C. 2008. The burden of vancomycin-resistant enterococcal infections in US hospitals, 2003 to 2004. *Diagn. Infect. Dis.* 62: 81-85.

Rende-Fournier, R.; Leclercq, R.; Galimand, M.; Duval, J. & Courvalin, P. 1993. Identification of the *satA* gene encoding a streptogramin A acetyltransferase in *Enterococcus faecium* BM4281. *Gene* 37: 2119-2125.

Reynolds, P.E. & Courvalin, P. 2005. Vancomycin resistance in enterococci due to synthesis of precursors terminating in D-alanyl-D-serine. *Antimicrob. Agents Chemother.* 49: 21-25.

Ribas, R.M.; Darini, A.L.; Moreira, T.A.; Freitas, C. & Gontijo Filho PP. 2007. Vancomycin-resistant VanA phenotype Enterococcus faecalis: first case in Minas Gerais state and epidemiological considerations. *Braz. J. Infect. Dis.* 11: 439-440.

Rice, L.B.; Carias, L.L.; Donskey, C.J. & Rudin, S.D. 1998. Transferable, plasmid-mediated VanB-type glycopeptide resistance in *Enterococcus faecium. Antimicrob. Agents Chemother.* 42: 963-964.

Rich, R.L.; Demeler, B.; Kreikemeyer, B.; Owens, R.T.; Labrenz, S.; Narayanna, S.V.; Weinstock, J.M.; Murray, B.E. & Hook, M. 1999. Ace is a collagen-binding MSCRAMM from *Enterococcus faecalis. J. Biol. Chem.* 274: 26939-26945.

Richards, M.J.; Edwards, J.R.; Culver, D.H. & GAYNES, R.P. 2000. Nosocomial infections in combined medical-surgical intensive care units in the United States. *Infect. Control. Hosp. Epidemiol.* 21: 510-515.

Roberts, M.C. & Hillier, L. 1990. Genetic basis of tetracycline resistance in urogenital bacteria. *Antimicrob. Agents Chemother.* 34 476-478.

Roberts, M.C.; Sutcliffe, J.; Courvalin, P.; Bogo Jensen, L.; Rood, J. & Seppala. 1999. Nomenclature for macrolide and macrolide-lincosamide-streptogramin B resistance determinants. *Antimicrob. Agents. Chemother.* 43: 2823-2830.

Rodríguez-Baño, J.; Ramírez, E.; Muniain, M.A.; Santos, J.; Joyanes, P.; González, F.; García-Sánchez, M. & Martinez-Martinez, L. 2005. Colonization by high-level aminoglycoside-resistant enterococci in intensive care unit patients: epidemiology and clinical relevante. *J. Hosp. Infect.* 60: 353-359.

Rosin, C.; Bernsmeier, C.; Entenza, J.M.; Moreillon, P., Frei, R.; Weisser, M.; Flückiger, U. 2012. Daptomycin for highly resistant *Enterococcus faecium* infection. *Swiss. Med.* Wkly 142: w13603.

Rossi, F. & Andreazzi, D. 2006. Overview of tigecycline and its role in the era of antibiotic resistance. *Braz. J. Infect. Dis.* 10: 203-216.

Rouff, K.L.; De La Maza, L.; Murtagh, M.J.; Spargo, J.D. & Ferraro, M.J. 1990. Species identities of enterococci isolated from clinical specimens. *J. Clin. Microbiol.* 28: 435-437.

Ruiz-Garbajosa, P.; Cantón, R.; Pintado, V.; Coque, T.M.; Willems, R.; Baquero. F. Del Campo, R. 2006. Genetic and phenotypic differences among *Enterococcus faecalis* clones from intestinal colonization and invasive disease. *Clin. Microbiol. Infect.* 12: 1193-1198.

Rybkine, T.; Mainardi, J.L.; Sougakoff, W.; Collatz, E. & GUTMANN, L. 1998. Penicillin-binding.protein 5 sequence alterations in clinical isolates of *Enterococcus faecium* with different levels of β-lactam resistance. J. *Infect. Dis.* 178: 129-163.

Sader, H.S.; Jones, R.N.; Gales, A.C.; Silva, J.B.; Pignatari, A.C. & The Sentry Participants Group (Latin America). 2004. SENTRY antimicrobial surveillance program report: Latin American and Brazilian results from 1997 through 2001. *Braz. J. Infect. Dis.* 8: 25-79.

Sader, H.S.; Watters, A.A.; Fritsche, T.R. & Jones, R.N. 2007. Daptomycin antimicrobial activity tested against methicillin-resistant staphylococci and vancomycin-resistant enterococci isolated in European medical centers (2005). *BMC Infect. Dis.* 18: 7-29.

Saeedi, B.; Hällgren, A.; Isaksson, B.; Jonasson, J.; Nilsson, L.E. & Hanberger. H. 2004. Genetic relatedness of *Enterococcus faecalis* isolates with high-level gentamicina resistance from patients with bacteraemia in the south east of Sweden 1994-20012. *Scand. J. Infect. Dis.* 36: 405-409.

Sahm, D.F.; Free, Smith, C.; Eveland, M. & Mundy, L.M. 1997. Rapid characterization schemes for surveillance isolates of vancomycin-resistant enterococci. *J. Clin. Microbiol.* 35: 2026-30.

Satake, S.; Clark, N.; Rimland, D.; Nolte, F.S. & Tenover, F.C. 1997. Detection of vancomycin-resistant enterococci in fecal samples by PCR. *J. Clin. Microbiol.* 35: 2325-2230.

Sava, I.G.; Heikens, E.; Huebner, J. 2010. Pathogenesis and immunity in enterococcal infections. *Clin. Microbiol. Rev.* 16: 533-540.

Schaberg, D.R.; Culver, D.H., & Gaynes, R.P. 1991. Major trends in the microbial etiology of nosocomial infections. *Am. J. Med.* 91: 79S-82S.

Scheetz, M.H.; Qi, C.; Noskin, G.A.; Warren, J.R.; Postelnick, M.J.; Malxzynski, M. Huang, J. & Zembower, T.R. 2006. The clinical impact of linezolid susceptibility reporting in patients with vancomycin-resistant enterococci. Diagn. *Microbiol. Infect. Dis.* 56: 407-413.

Scheetz, M.H.; Knechtel, S.A., Malczynski, M. Postelnick, M.J. & QI, C. 2008. Increasing incidence of linezolide-intermediate or –resistant vancomycin-resistant *Enterococcus faecium* strains parallels increasing linezolid consumption. *Antimicrobi. Agents Chemother.* 52: 2256-2259.

Schleifer, K.H. & Kilpper-Bälz, R. 1984. Transfer of *Streptococcus faecalis* and *Streptococcus faecium* to the genus *Enterococcus* nom. rev. as *Enterococcus faecalis* comb. nov. and *Enterococcus faecium* comb. nov. *Int. J. Syst. Bacteriol.* 34: 31-34.

Schoonmaker, D.J.; Bopp, L.H.; Baltch, A.; Smith, R.P.; Rafferty, M.E. & George, M. 1998. Genetic analysis of multiple vancomycin-resistant *Enterococcus* isolates obtained serially from two long-term-care patients. *J. Clin. Microbiol.* 36: 2105- 2108.

Schulte, B.; Heininger, A.; Autenrieth, I.B. & Wolz, C. 2008. Emergenge of increasing linezolid-resistance in enterococci in a post-outbreak situation with vancomycin-resistant *Enterococcus faecium*. *Epidemiol. Infect.* 136: 1131-1133.

Shankar, N., Lockatell, C.V., Baghdayan, A.S., Drachenberg, C., Gilmore, M.S. & Johnson, D.E. 2001. Role of *Enterococcus faecalis* surface protein Esp in the pathogenesis of ascending urinary tract infection. *Infect. Immun.*, 69: 4366-4372.

Shepard, B.D. & Gilmore, M.S. 2002. Antibiotic-resistant enterococci: the mechanisms and dynamics of drug introduction and resistance. *Microbes* 4: 215-224.

Singh, K.V.; Coque, T.M.; Weinstock, G.M. & Murray, B.E. 1998. In vivo testing of an *Enterococcus faecalis* efaA mutant and use of efaA homologs for species identification. FEMS Immunol. *Med. Microbiol.* 21: 323-331.

Soltani, M.; Beighton, D.; Philpott-Howard, J. & Woodford, N. 2000. Mechanisms of resistance to quinupristin-dalfopristin among isolates of *Enterococcus faecium* from animals, raw meat, and hospitals patients in Western Europe. *Antimicrob. Agents Chemother.* 44: 433-436.

Song, J. Ko, K.S.; Suh, Y.; Oh, W.S. ; Kang, C. ; Chung, D.R.; Peck, K.R.; Lee, N.Y. & Lee, W.G. 2008. Clinical implications of vancomycin-resistant *Enterococcus faecium* (VRE) with VanD phenotype and *vanA* genotype. *J. Antimicrob. Chemother.* 61: 838-844.

Sung, K.; Khan, S.A. & Nawaz, M.S. 2008. Genetic diversity of Tn1546-like elements in clinical isolates of vancomycin-resistant enterococci. *Int. J. Antimicrob. Agents* 31: 549-554.

Stern, C.S.; Carvalho, M.G.S. & Teixeira, L.M.1994. Characterization of enterococci isolated from human and nonhuman sources in Brazil. *Diagn. Microbiol. Infect. Dis.*, 20: 61-67.

Stratton, C.W. & Cooksey, R.C. 1991. Susceptibility tests: special tests. *In*: Ballows, A., Hausler, W. J., Jr., Herman, K. L., Isenberg, H. D. and Shadomy, H. J. Manual of Clinical Microbiology. 5th ed. American Society for Microbiology. Washington, D.C. p: 1153-1165

Struelens M. J. & the Members of the European Study Group on Epidemiological Markers (ESGEM) of the European Society for Clinical Microbiology and Infection Diseases (ESCMID). 1996. Consensus guidelines for appropriate use and evaluation of microbial epidemiologic typing systems. *Clin. Microbiol.* Infect. 2: 2-11.

Sujatha, S.; Praharaj, I. 2012. Glycopeptide resistance in gram-positive cocci: a review. *Interdiscip. Perspect. Infect. Dis.* 2012: 1-10.

Sussmuth, S.D.; Muscholl-Silberhorn, A.; Wirth, R.; Susa, M.; Marre, R. & Rozdinski, E. 2000. Aggregation substance promotes adherence, phagocytosis, and intracellular survival of *Enterococcus faecalis* with human macrophages and suppresses respiratory burst. *Infest Immun.* 68: 4900-4906.

Sutcliffe, J.; Tait-Kamradt, A. & Wondrack, L. 1996. *Streptococcus pneumoniae* and *Streptococcus pyogenes* resistant to macrolides but sensitive to clindamycin: a common resistance pattern mediated by efflux system. *Antimicrob. Agents Chemother.* 40: 1817-1824.

Svec, P.; Devriese, L.A.; Sedlacek, I.; Baele, M.; Vancanneyt, M.; Haesbrouck, F.; Swings, J. & Doskar, J. 2001. *Enterococcus haemoperoxidus* sp. Nov. And *Enterococcus moraviensis* sp. Nov., isolated from water. *Int. J. Syst. Evol. Microbiol.*, 51: 1567-1574.

Svec, J.M.; Vancanneyt, M.; Koort, J.; Naser, S.M.; Hoste, B.; Vihavainen, E.; Vandamme, P.; Swing, J. & Björkroth, J. 2005. *Enterococcus devriesei* sp. nov., associated with animal sources. *Int. J. Syst. Evol. Microb.* 55: 2479-2484.

Svec, J.M.; Vancanneyt, M.; Sedlácek, I.; Naser, S.M.; Snauwaert, C.; Lefebvre, K.; Hoste, B. & Swing, J. 2006. *Enterococcus silesiacus* sp. nov. and *Enterococcus termitis* sp. nov. *Int. J. Syst. Evol. Microb.* 56: 577-581.

Swenson, J.M.; Ferraro, M.J.; Sham, D.F.; Clarck, N.C.; Culver, D.H.; Tenover, F.C. & The National Committee For Clinical Laboratory Standard Working Group on Enterococci. 1995. Multilaboratory evaluation of screening methods for detection of high-level aminoglycoside resistance in enterococci. *J. Clin. Microbiol.* 33: 3008-3018.

Tait-Kamradt, A.; Clancy, J.; Cronan, M.; Dib-Hsjj, F.; Wondrack, L.; Yuan, W. & Sutcliffe, J. 1997. *mefE* is necessary for erytromycin-resistant M phenotype in *Streptococcus pneumoniae*. *Antimicrob. Agents Chemother.* 41: 2251-2255.

Tankovic, J.; Mahjoubi, F.; Courvalin, P.; Duval, J. & Leclercq, R. 1996. Development of fluoroquinolone resistance in *Enterococcus faecalis* and role of mutations in the DNA gyrase gyrA gene. *Antimicrob. Agents Chemother.* 40:2558-2561.

Teixeira, L.M.; Facklam, R.R.; Steigerwalt, A.G.; Pigott, N.E.; Merquior, V.L.C. & Brenner, D.J. 1995. Correlation between phenotypic characteristics and DNA relatedness within *Enterococcus faecium* strains. *J. Clin. Microbiol.* 33: 1520-1523.

Teixeira, L.M.; Carvalho, M.G.S.; Merquior, V.L.C.; Steigerwalt, A.G.; Brenner, D. J. & Facklam, R.R. 1997a. Phenotypic and genotypic characterization of *Vagococcus fluvialis*, including strains isolated from human sources. *J. Clin. Microbiol.* 35: 2778-2781.

Teixeira, L.M.; Carvalho, M.G.S.; Merquior, V.L.C.; Steigerwalt, A.G.; Teixeira, M. G.M.; Brenner, D.J. & Facklam, R.R. 1997b. Recent approaches on the taxonomy of the enterococci and some related microorganisms. *Adv. Exp. Med. Biol.* 418: 387-391.

Teixeira, L.M.; Paula, G.R.; Mondino, S.S.B.; Mendonça, C.R.; Carvalho, M.G.S.; D'azevedo, P.A.; Stern, C.S.; Stuckert, A.P., Castro, A.C. & Facklam, R.R. 2000. Phenotypic and genotypic characterization of enterococci isolated from human clinical sources during, 1985 to 1998, in Brazil. Abstracts of the 1st International ASM Conference on Enterococci: pathogenesis, biology, and antibiotic resistance. American Society for Microbiology, Washington, D.C. p.27.

Teixeira, L. M.; D'azevedo, P.; Dias, C. G.; Sukiennik, T.; Hentges, J. D.; Gonçalves, A. L.; Merquior, V. L. C. & Carvalho, M. G. S. 2001. Genetic relationship of vancomycin-resistant *Enterococus faecalis* strains carrying the *vanA* gene isolated in Porto Alegre City, Brazil. Apresentado no "Interscience Conference on Antimicrobial Agents and Chemotherapy".

Teixeira, L.M.; Carvalho, M.G.S.; Espinola, M.M.B.; Steigerwalt, A.G.; Douglas, M. P.; Brenner, D.J.& Facklam, R.R. 2001. *Enterococcus porcinus* sp. nov. and *Enterococcus ratti* sp. nov.: novel species associated with enteric disorders in animals. *Int. J. Syst. Evol. Microbiol.* 51: 1737-1743

Teixeira, L.M.; Carvalho, M.G. & Facklam RR. 2007. *Enterococcus*. *In*: Murray P, Baron EJ, Jorgensen JH, Landry ML, Pfaller MA. *Manual of Clinical Microbiology*, 9th ed., American Society for Microbiology, Washington, DC. p 430-442

Tenover, F.C.; Arbeit, R.D.; Georing, R.V.; Mickelsen, P.A.; Murray, B.E.; Persing, D.H. & Swaminathan, B. 1995. Interpreting chromosomal DNA restriction patterns produced by pulsed-field gel electrophoresis: criteria for bacterial strain typing. *J. Clin. Microbiol.* 33: 2233-2239.

Tenover, F.C.; Arbeit, R.D.; Goering, R.V. And Molecular Typing Working Group Of The Society for Healthcare Epidemiology of America. 1997. How to select and interpret molecular strain typing methods for epidemiological studies of bacterial infections: a review for healthcare epidemiologists. *Infect. Control. Hosp. Epidemiol.* 18: 426-439.

Titze-De-Almeida, R.; Rollo Filho, M.; Silveira, C.A.N.; Rodrigues, I.P.; Eudes Filho, J.; Nascimento, R.S.; Ferreira Ii, R.F.; Moraes, L.M.P.; Boelens, H.; Van Belkum, A. & Felipe, M.S.S. 2004. Molecular epidemiology and antimicrobial susceptibility of enterococci recovered from Brazilian intensive care units. *Braz. J. Infect. Dis.* 8: 197-205.

Titze-De-Almeida, R.; Van Belkum, A.; Felipe, M.S.S.; Zanella, R.C. Top, J. & Willems, R.J. 2006. Multilocus sequence typing of hospital-associated Enterococcus faecium from Brazil reveals their unique evolutionary history. *Microb. Drug Resist.* 12: 121.

Tomayko, J.F. & Murray, B.E. 1995. Analysis of *Enterococcus faecalis* isolates from intercontinental sources by multilocus enzyme electrophoresis and pulsed-field gel electrophoresis. *Antimicrob. Agents Chemother.* 33:2903-2907.

Tomita, H.; Pierson, C.; Lim, S.K.; Clewell, D.B. & Ike, Y. 2002. Possible connection between a widely disseminated conjugative gentamicin resistance (pMG1-like) plasmid and the emergence of vancomycin resistance in *Enterococcus faecium*. *J. Clin. Microbiol.* 40: 3326:3333.

Top, J.; Schouls, L.M.; Bonten, M.J. & Willems RJ. 2004. Multiple-locus variable-number tandem repeat analysis, a novel typing scheme to study the genetic relatedness and epidemiology of Enterococcus faecium isolates. *J. Clin. Microbiol.* 42:4503-4511.

Top, J.; Banga, N.M.; Willems, R.J.; Bonten, M.J. & Hayden, M.K. 2008. Comparison of multiple-locus variable-number tandem repeat analysis and pulsed-field gel electrophoresis in a setting of polyclonal endemicity of vancomycin-resistant *Enterococcus faecium*. *Clin. Microbiol. Infect.* 14: 363-369.

Top, J.; Willems, R.; van der Velden, S.; Asbroek, M. & Bonten, M. 2008. Emergence of Clonal Complex 17 *Enterococcus faecium* in The Netherlands. *J. Clin. Microbiol.* 46: 214-219.

Trallero, E.P.; Urbieta, M.; Montes, M.; Ayestaran, I. & Marinom JM. 1998. Emergence of *Streptococcus pyogenes* strains resistant to erythromycin in Gipuzkoa, Spain. Eur J. Clin. Microbiol. Infect. Dis. 17: 25-31.

Tressoldi, A.T.; Cardoso, L.G.; Castilho, G.V.; Dantas, S.R.; Vonnowakonski, A.; Pereira, R.M. & Trabasso, P. 2006. Low prevalence of vancomycin resistant enterococci colonization in intensive care patients in a Brazilian teaching hospital. *Braz. J. Infect. Dis.* 10: 239-241.

Trzcinski, K.; Cooper, B.S.; Hryniewicz, W. & Dowson, C.G. 2000. Expression of resistance to tetracyclines in strains of methicillin-resistant *Staphylococcus aureus*. *J. Antimicrbol. Chemother.*; 45: 763-770.

Tyrrell, G.J.; Turnbull, L.; Teixeiram L,M.; Lefebvrem J,. Carvalho, M.G. & Facklam Rr, Lovgren M. 2002. Enterococcus gilvus sp. nov. and Enterococcus pallens sp. nov. isolated from human clinical specimens. *J. Clin. Microbiol.* 4: 1140-1145.

Umgelter, A.; Prinz, C.; Gaa, J. & Huber, W. 2007. Ascending pneumonia complicating endoscopic therapy of a pancreativ abscess. *Endoscopy* 39: 267.

Vakulenko, S.B.; Donabedian, S.M.; Voskresensky, A.M.; Zervos, M.J.; Lerner, S.A. & Chow, J. W. 2003. Multiplex PCR for detection of aminoglycoside resistance genes in enterococci. *Antimicrob. Agents Chemother.* 47: 1423-26.

Van Belkum, A.; Van Leeuwen, W.; Kluytmans, J. & Verbrugh, H. 1995. Molecular nosocomial epidemiology: high speed typing of microbial pathogens by arbitrary primed polymerase chain reaction assays. *Infect. Control Hosp. Epidemiol.* 16: 658-666.

Vancanneyt, M.; Valderato, F.; Loygren, M.; Tyrrel, G.J.; Bokermann, S.; Almeida, S.C.G.; Descheemaeker, P.; Goossens, H.; Pot, B.; Vandamme A.; Swings, J.; Haesebrouck, F. & Devriese, L.A. 2001. *Enterococcus villorum* sp. nov., an enteroadherent bacterium associated with diarrhea in piglets. *Int. J. Systematic Evol.. Microbiol.* 51: 393-400.

Vilela, M.A.; Souza, S.L.; Palazzo, I.C.; Ferreira, J.C.; Morais, M.A.; Darini, A.L. & Morais, M.M. 2006. Identification and molecular characterization of *vanA*-type vancomycin-resistant *Enterococcus faecalis* in Northeast of Brazil. *Mem. Inst. Oswaldo Cruz* 101: 715-719.

Xu, X.; Lin, D.; Yan, G.; Ye, X.; Wu, S.; Guo, Y.; Zhu, D.; Zhang, Y.; Wang, F.; Jacoby, G.A.; Wang, M. 2010. vanM, a new glycopeptides resistance gene cluster found in *Enterococcus faecium*. *Antimicrob. Agents Chemother.*, 54: 4643-4647.

Wade, J.J. 1997. *Enterococcus faecium* in hospitals. *Eur. J. Clin. Microbiol. Infect. Dis.* 16: 113-119.
Weaver, K.E.; Rice, L.B. & Churchward, G. 2002. Plasmid and transposons. *In:* Gilmore, M.S., Clewell, D.B., Courvalin, P., Dunny, G.M., Murray, B.E. & Rice, L.B. *The enterococci: pathogenesis, molecular biology, and antibiotic resistance.* ASM Press, Washington DC. p 219-263.
Weisblum, B. 1995. Erytromycin resistance by ribossome modification. *Antimicrob. Agents Chemother.* 39: 577-585.
Werner, G.; Klare, I.; Heier, H.; Hinz, K.H.; Bohme, G.; Wendt,M. & Witte, W. 2000. Quinupristin/dalfopristin-resistant enterococci of the *satA* (*vatD*) and *satG* (*vatE*) genotypes from different ecological origins in Germany. *Microb. Drug Res.* 6: 37-47.
Werner, G.; Gförer, S.; Fleige, C.; Witte, W & Klare, I. 2008. Tigecycline-resistant *Enterococcus faecalis* strain isolated from a German intensive care unit patient. *J. Antimicrob. Chemother.* 61: 1182-1183.
Willey, B.M.; Mcgeer, A.J.; Ostrowski, M.A.; Kreiswirth, B.N. & Low, D.E. 1994. The use of molecular typing techniques in the epidemiologic investigation of resistant enterococci. *Infect. Control. Hosp. Epidemiol.* 15: 548-556.
Willems, R.J.; Hanage, W.P.; Bessen, D.E.; Feil, E.J. 2011. Population biology of Gram-positive pathogens: high-risk clones for dissemination of antibiotic resistance. *FEMS Microbiol. Rev.* 35: 872-900.
Willems, R.J.; Top, J.; Van Santen, M.; Robinson, D.A.; Coque, T.M.; Baquero, F.; Grundmann, H. & Bonten, M.J. 2005. Global spread of vancomycin-resistant *Enterococcus faecium* from distinct nosocomial genetic complex. *Emerg. Infect. Dis* 11: 821-828.
Wickman, P.A.; Black, J.A.; Moland, S.; Thomson, K.S. & Hanson, N.D. 2006. In vitro development of resistance to DX-619 and other quinolones in enterococci. *J. Antimicrob. Chemother.* 58: 1268-1273.
Yazgi, H.; Ertek, M.; Erol, S. & Ayyildiz, A. 2002. A comparison of high-level amonoglycoside resistance in vancomycin-sensitive and vancomycin-resistant *Enterococcus spp*ecies. *J. Int. Medical Research* 30: 529-534.
Zanella, R.C.; Valdetaro, F.; Lovgren, M.; Tyrrel, G.J.; Bokermann, S.; Almeida, S.C.G.; Vieira, V.S.D. & BRANDILEONE, M.C.C. 1999. First confirmed case of a vancomycin-resistant *Enterococcus faecium* with *van*A phenotype from Brazil: isolation from a meningitis case in São *Paulo. Microb. Drug Res.* 2: 159-162.
Zanella, R.C.; Brandileone, M.C.; Bokermann, S.; Almeida, S.C.; Valdetaro, F.; Vitório, F.; Moreira Mde, F.; Villins, M.; Salomão, R. & Pignatari, A.C. 2003. Phenotypic and genotypic characterization of VanA *Enterococcus* isolated during the first nosocomial outbreak in Brazil. *Microb. Drug Resist.* 9: 283-291.
Zarrilli, R.; Tripoldi, M.F.; Di Popolo, A.; Fortunato, R.; Bagattini, M.; Florio, A. Triassi, M.; Utili, R. 2005. Molecular epidemiology of high-level amonoglycoside-resistant enterococci isolated from patients in a university hospital in southern Italy. *J. Antimicrob. Chemother.* 56: 827-835.
Zervos, M.J., Kauffman, C.A., Therasse, P.M., Bergman, A.G.T., Mikesell, Schaberg, D.R. 1987. Nosocomial infection by gentamicin-resistant *Streptococcus faecalis*: an epidemiologic study. *Ann. Intern. Med.* 106: 687-69.

Zhanel, G.G.; Decorby, M.; Laing, N.; Weshnoweski, B.; Vashiht, R.; Tailor, F.; Nochol, K.A.; Wierzbowski, A.; Baudry, P.J.; Karlowsky, J.A.; Lagacé-Wiens, P.; Walkty, A.; Mccracken, M.; Mulvey, M.R.; Johnson, J.; Canadian Antimicrobial Resistance Alliance (Cara) & Hoban, D.J. 2008. Antimicrobial-resistant pathogens in intensive care units in Canada: results of the Canadian National Intensive Care Unit (CAN-ICU) study, 2005-2006. *Antimicrob. Agents Chemother.* 52: 1430-1437.

Zheng, B.; Tomita, H.; Xiao, Y.H.; Wang, S.; LI, Y. & IKE, Y. 2007. Molecular characterization of vancomycin-resistant *Enterococcus faecium* isolates from Mainland China. *J. Clin. Microbiol.* 45: 2813-2818.

Zirakzadeh, A. & Patel, R. 2006. Vancomycin-resistant enterococci: colonization, infection, detection, and treatment. *Mayo Clin. Proc.* 81: 529-536.

Zscheck, K.K., & Murray, B.E. 1993. Genes involved in the regulation of β-lactamase production in enterococci and staphylococci. *Antimicrob. Agents Chemother.*, 37: 1966-1970.

In: *Enterococcus faecalis*
Editor: Henry L. Mack

ISBN: 978-1-63321-049-3
© 2014 Nova Science Publishers, Inc.

Chapter 4

PROMISCUITY, PHEROMONES AND PATHOGENICITY: WHY ALL ENTEROCOCCI ARE NOT CREATED EQUAL

Elise Pelzer (PhD)[1,2], Irani Rathnayake (PhD)[2] and Flavia Huygens (PhD)[2]

[1]The Wesley Research Institute, Women's Health Laboratory,
The Wesley Hospital, Auchenflower, Queensland, Australia
[2]School of Biomedical Sciences, Institute of Health
and Biomedical Innovation, Faculty of Health,
Queensland University of Technology, Brisbane, Queensland, Australia

ABSTRACT

Enterococcus faecalis is a Gram-positive, coccus shaped, lactic acid bacterium, with demonstrated ubiquity across multiple anatomical sites. *Enterococcus faecalis* isolates have been isolated from clinical samples as the etiological agent in patients with overt infections, and from body sites previously thought to be sterile but absent of signs and symptoms of infection.

E. faecalis is implicated in both human health and disease, recognized as a commensal, a probiotic and an opportunistic multiply resistant pathogen. *E. faecalis* has emerged as a key pathogen in nosocomial infections.

E. faecalis is well equipped to avert recognition by host cell immune mediators. Antigenic cell wall components including lipotechoic acids are concealed from immune detection by capsular polysaccharides produced by some strains. Thereby preventing complement activation, the pro-inflammatory response, opsonisation and phagocytosis. *E. faecalis* also produces a suite of enzymes including gelatinase and cytolysin, which aid in both virulence and host immune evasion. The ability of enterococci to form biofilms *in vivo* further increases virulence, whilst simultaneously preventing detection by host cells.

E. faecalis exhibits high levels of both intrinsic and acquired antimicrobial resistance. The mobility of the *E. faecalis* genome is a significant contributor to antimicrobial resistance, with this species also transferring resistance to other Gram-positive bacteria.

Whilst *E. faecalis* is of increasing concern in nosocomial infections, its role as a member of the endogenous microbiota cannot be underestimated. As a commensal and probiotic, *E. faecalis* plays an integral role in modulating the immune response, and in providing endogenous antimicrobial activity to enhance exclusion or inhibition of opportunistic pathogens in certain anatomical niches.

In this chapter we will review possible mediators of enterococcal transition from commensal microbe to opportunistic pathogen, considering isolates obtained from patients diagnosed with pathogenic infections and those obtained from asymptomatic patients.

1. *E. FAECALIS* EPIDEMIOLOGY

E. faecalis, is a ubiquitous and robust bacterium, noted for its ability to transition from harmless commensal to opportunistic pathogen in clinical environments [1]. The key to the dualistic role of *E. faecalis* as both a commensal and an opportunistic pathogen appear related to the molecular mechanisms engaged upon interactions with the host; however, there is as yet no clear cut method to differentiate between avirulent and virulent strains [2]. If the host-microbe balance is disrupted in any way, then *E. faecalis* actively alters its growth in response to environmental cues, and seeks to acquire virulence traits to facilitate infection of host cells [3]. In order to understand the transition from commensal to pathogen, one must consider how virulence traits are inherited and the regulatory mechanisms associated with such traits in virulent strains.

The central mechanism in enterococcal transition from commensal to pathogen appears to be related to genetic promiscuity in this species. A large proportion of mobile elements within the genome contribute to antimicrobial resistance, and the spread of virulence traits particularly via pathogenicity islands (PAI), which represent gene clusters encoding proteins with known roles in transcriptional regulation, gene transfer, virulence, and stress survival [4].

2. IMMUNE EVASION/IMMUNOMODULATION

Lipoteichoic acids (LTA) are surface associated amphiphilic adhesion molecules in Gram-positive bacteria that regulate autolytic cell wall enzymes. LTAs, which reportedly comprise up to 60% of the Gram-positive bacterial cell wall, are key contributors in activating the host immune response and generating the overall net charge of the bacterial cell envelope [5]. As with other Gram-positive bacteria, the immunostimulatory potential of *E. faecalis* is intimately linked to the cell wall LTA composition, which results in differential innate immune activation and Toll-like receptor activation potency [6]. Cell wall LTA composition is a determinant of pathogenicity. The surface exposed bacterial lipoproteins are versatile immunomodulators, contributing to the host immune response to invasion via regulation, recruitment and activation of leukocytes during inflammation and the resolution of infection. LTA binds to TLR-2, inducing intracellular messengers (MYD88, TRAF6 and MAP kinases), which activate the transcription factor signalling pathways (NFκβ and AP-1) essential for inducing the production and secretion of pro-inflammatory cytokines [7,8]. During infection, *E. faecalis* LTA activates phagocytic polymorphonuclear neutrophil granulocytes (PMNs) resulting in cell elongation, degranulation priming for oxidative burst and production of specific pro-inflammatory cytokines including interleukin (IL)-8, tumour necrosis factor (TNF)-α and granulocyte colony stimulating factor (G-CSF) [9]. LTA also stimulates cytokine production by monocytes, which release IL-1β, IL-6, TNF-α and the chemokine IL-8, which further recruit PMNs. Pro-inflammatory agonists appear to be enhanced by LTA and peptidoglycan synergy [10]. In some instances, LTA appears to inhibit spontaneous apoptosis in PMNS, thereby increasing their lifespan, which can improve infection resolution, but as a consequence, can prolong inflammation and damage host cells [6]. Collectively these lead to bacterial and host cell damage. LTA reportedly results in differential activation of the host innate immune response to Gram-positive bacteria. *In vitro* studies of commensal lactobacilli report strain specific differences in chemokines and cytokine activation, likely related to host cell damage. Similar Gram-positive commensal species, with high colonizing capacity such as *E. faecalis* would be most beneficial to the host if they elicited only low level inflammation and thus remained adherent to host cells providing physical and chemical barriers to adherence and invasion by pathogens.

Damage to the bacterial cell wall by the host immune response, release of leukocyte cationic peptides, lysozyme induced bacteriolysis, or treatment with β-lactam antibiotics can all liberate LTA [11,12]. The bacterial LTA facilitates both specific and non-specific cell binding via membrane phospholipids. Specific cell binding occurs via CD14 and TLR cell surface receptors, whilst non-specific binding occurs via binding to host cell fibronectin, albumin or fatty acids in albumin [13]. Ligand binding promotes interaction with circulating antibodies or complement inducing opsonophagocytosis, or neutrophil and macrophage release of reactive oxygen and/or nitrogen species, proteases, hydrolases, bactericidal cationic peptides, growth factors and pro-inflammatory cytokines [14]. *E. faecalis* LTA and diheteroglycan are the targets of host opsonic antibodies [15]. However, on encapsulated strains, key antigenic targets of the opsonic antibodies and subsequent complement cascade are masked by capsule polysaccharides [16].

3. VIRULENCE MECHANISMS

E. faecalis has emerged as a serious opportunistic pathogen in modern medical science. This commensal turned pathogen is highly adaptable and can withstand harsh environmental conditions, is able to tolerate high salt concentrations and a broad temperature range [17]. The success of *E. faecalis* as an opportunistic pathogen is in part due to its ability to adapt to, and survive normally lethal conditions following exposure to sub-lethal stress [17]. In response to stress, this species enters a viable but non-cultivable state, capable of resuscitation when favourable conditions return [18]. *E. faecalis* has two alternate survival stress responses. As a non-sporulating species, *E. faecalis* can activate a viable but non-cultivable state to persist as vegetative cells with low metabolic activity under hostile environmental conditions. In contrast, where nutrient supply is restricted or exhausted, a starvation response is activated, producing cells which persist long-term in a non-growing but cultivable state [19]. In addition to its inherent ability to survive adversity, *E. faecalis* produces a suite of virulence factors, which function to enhance infection and survival in the host environment, which have been reviewed elsewhere (Summarized in Table 1). Virulence factors such as gelatinase, esp, and biofilm formation are not absent from commensal *E. faecalis* strains, but are reportedly less common, suggesting that these factors potentially play a role in maintaining commensal stability within the host, not just in pathogenesis [20,21]. Table 2 represents data from *E. faecalis* urogenital tract isolates of clinical origin collected from public and private hospitals in Brisbane, Australia. None of the isolates harboured the resistance genes for *vanA, vanB* or *tetS*, however, *esp* and *tetM* genes were present in the majority of commensal and clinical isolates. Virulence genes including *cyl-A, cyl-B, cyl-M*, and the antibiotic resistance genes *gyrA* and *aac(6')-aph(2')* were detected only in *E. faecalis* isolated from urine samples but not genital tract sites. Detection of antibiotic reisitance in our isolates was similar to that reported by other groups [22]. We detected resistance genes for gentamicin, ciprofloxacin and tetracycline in our clinical isolates but only tetracycline resistance genes in commensal isolates of genital tract origin. We identified a single ampicillin resistant isolate, likely representing acquired resistance given that ampicillin resistance is not always encountered in clinical *E. faecalis* isolates [23-25]. Previous studies employing commensal and pathogenic *E. faecalis* strains in models of human infection reported the presence and fitness of virulence factors, as well as the temporal regulation in response to the host environment as defining factors in pathogenesis rather than the expression of such traits [26-28]. Differential expression of virulence related genes in *E. faecalis* has previously been reported in response to biological cues in urine, possibly supporting the increased detection of these genes in the urinary isolates in our study [29]. Others have also identified the presence of a combination of virulence factors including *gel-E*, *esp* and haemolysin genes in urines obtained from clinical infections, with factors identified in urinary isolates deemed to be of higher virulence potency [22,24,30-32]. Within our samples, we identified a number of virulence profiles with *E. faecalis* isolates elaborating both virulence genes and antibiotic resistance genes. The identification of all virulence and antibiotic resistance genes in *E. faecalis* isolates from our study was higher in the clinical compared to the commensal isolates. Isolates of genital tract origin (29%) produced a single profile of *esp* and *TetM* compared to the urinary isolates of which a proportion produced the same *esp* and *tetM* profile (9%) but of which multiple others produced two additional profiles (12% and 9% respectively): *cyl-B, cyl-M, esp* and *tetM*; and *cyl-B, cyl-M, gel-E, gyrA* and *aac(6')-aph(2')*, *tetL* and *tetM*. Taken together, these results support the notion that the transition from commensal to pathogen may actually be a continuum supported by the promiscuous *E. faecalis* genome, which demonstrates an exceptional capacity to evolve to suit the niche environment through acquisition of virulence factors improving fitness and survival.

Virulence factor genes have been detected in both commensal and pathogenic *E. faecalis*, at comparable rates for *esp*, and *fsrb*. Biofilm formation also reportedly occurred at similar rates in both cohorts with no significant difference reported for any groups when comparing *esp*-positive and *esp*-negative isolates or *fsrb*-positive and *fsrb*-negative isolates [20]. Commensal and pathogenic *E. faecalis* isolates also elaborated similar biochemical profiles for hemolysin and gelatinase production [20]. Hemolysin production did not correlate with biofilm formation in commensal isolates; however, haemolytic pathogenic isolates produced a higher mean biofilm value when compared to non-hemolytic isolates. In contrast, gelatinase producing commensal isolates produced biofilms of higher mean values than did gelatinase-negative commensal strains, whilst no difference was observed in the mean biofilm value for gelatinase-positive compared to gelatinase-negative pathogenic *E. faecalis* isolates [20]. Investigations of *E. faecalis* adhesion properties, which focused on aggregation substance, enterococcal surface protein and cell surface charge have revealed an increased heterogeneity in cell surface charge in pathogenic versus commensal isolates, which was independent of the presence of aggregation substance, enterococcal surface protein, culture conditions, growth phase and quorum sensing [33]. Population heterogeneity and cell surface charge may therefore be contributing factors in enterococcal pathogenesis.

4. SIGNALLING, QUORUM SENSING AND GENE TRANSFER

Cell-cell signalling is used to control density-dependent behaviour in bacteria. For Gram-positive species, cell-cell signalling or quorum sensing usually occurs via peptide mediated intercellular signalling [34]. Quorum sensing in bacteria represents communication between bacteria, whereby each species is able to determine information related to their population density in order to respond as a population in a coordinated manner [35]. The steps involved in the process include the secretion of specific signalling molecules by each cell, an accumulation of secretory molecules, followed by threshold concentration sensing leading to the induction of regulatory signalling cascades [36]. Quorum sensing plays a major role in virulence factor production and in biofilm formation.

Biofilms are surface attached bacterial communities contained within an extracellular matrix comprised of carbohydrates, secreted proteins and extracellular DNA [35]. Similar to other Gram-positive bacterial species, *E. faecalis* biofilm formation is regulated by carbohydrate metabolism [67]. Isolates expressing extracellular polysaccharide survive for longer periods within macrophages than non-polysaccharide producers [68]. Extracellular DNA is also produced at early time points by *E. faecalis* biofilms independently of cell lysis, and appears to represent a significant factor in biofilm stabilization [69]. The extracellular DNA is produced by a subpopulation of competent cells within a population of metabolically inactive or non-cultivable cells. Adhesion and biofilm formation is caused by cell surface charge and hydrophobicity [47,70-72]. More often than not, biofilm formation in bacterial communities requires activation of density-dependent gene expression mediated through cell-cell signalling and quorum sensing [73]. The established systems enterococci have evolved for sensing and responding to extracellular signals undoubtedly contribute to their robust phenotype and ability to survive under adverse conditions. Biofilm associated cells behave in a very different manner to planktonic cells as a result of community cell-cell signalling. Biofilm phenotypes demonstrate reduced growth rates, and increased antibiotic resistance and gene transfer [74-78].

Interestingly, screening of vancomycin-resistant and vancomycin-sensitive isolates collected from colonized and infected hospital patients, and from fecal samples from healthy volunteers elaborated a co-expression profile of the enterococcal virulence factors: gelatinase, aggregation substance and enterococcal surface protein common to both cohorts [79]. These factors have each been implicated in biofilm formation, and as such may represent a significant link between commensal and pathogen. *E. faecalis* isolates obtained from clinical and fecal samples also demonstrated similar expression of *esp* and gelatinase, as well as an equivalent propensity to form biofilms, again highlighting the potential of this species to transition from commensal to pathogen via dissemination and acquisition of virulence factors by promoting cell-cell interaction [20].

Table 1. *E. faecalis* virulence factors and their characterized functions

Virulence factor	Abbreviation	Function	Location	References
Adhesion of collagen	Ace	▪ Mediates adherence to collagen (type I and IV) and laminin	Cell surface	[37]
Aggregation substance	AS	▪ Bacterial cell clumping ▪ Plasmid-associated transfer and survival in neutrophils and internalization by epithelial cells	Cell surface Pheromone inducible conjugative plasmid	[38]
Capsular polysaccharide	cps	▪ Resistance to complement ▪ Resistance to phagocytosis and killing by PMNs ▪ Cytokine production	Cell surface	[39]
Cell wall carbohydrate		▪ Pili and biofilm formation	Cell surface	[40-42]
Endocarditis and biofilm associate pili	Ebp	▪ Mediate adherence to host cells: fibrinogen, collagen	Cell surface	[40-42]
Enterococcal leucine rich repeat containing protein	ElrA	▪ Macrophage infection	Cell surface	[43]
Enterococcal polysaccharide antigen	epa	▪ Biofilm formation and susceptibility to PMN phagocytosis and killing	Cell surface	[44,45]
Enterococcal surface protein	esp	▪ Role in biofilm formation	Cell surface Pathogenicity island	[46]
Glycolipids		▪ Biofilm formation	Cell surface	[47,48]
Lipoteichoic acid	LTA	▪ Facilitates binding to host cells ▪ Target for opsonic antibodies ▪ Promotes pro-inflammatory cytokine production by host cells	Cell surface	[49-52]

Virulence factor	Abbreviation	Function	Location	References
Microbial surface components recognizing adhesive matrix molecules	MSCRAMM	■ Bind components of host cell extracellular matrix	Cell surface	[37]
Pili		■ Adhesion to host cells ■ Biofilm formation	Cell surface	[40]
Sortase A	SrtA	■ Biofilm formation	Cell surface	[53]
Surface adhesins		■ Adhesion to host cells	Cell surface	[54] [55]
Wall techoic acid	WTA	■ Masks lecithin cell surface receptors to mediate complement activation	Cell surface	
Pathogenicity island encoded regulator	PerA	■ Biofilm formation ■ Survival in macrophages	Other	[37]
Thiol peroxidase	Tpx	■ Protection within phagocytic cells	Other	[56]
Bacteriocins		■ Inhibition of competing bacterial species	Secreted	[57]
Cytolysin	Cyl	■ Lysis of erythrocytes, PMNs and macrophages ■ Inhibits other species	Secreted	[37,58-60]
Extracellular superoxide		■ Host tissue damage	Secreted	[61]
Gelatinase	gelE	■ Biofilm formation ■ Host tissue damage	Secreted	[62,63]
Hyaluronidase		■ Promotes invasion of host cells by degrading hyaluronic acid ■ Bacterial nutrition	Secreted	[64]
Sex pheromones		■ Signalling peptide for plasmid acquisition ■ Inflammation in host	Secreted	[65,66]

Table 2. Prevalence of virulence and antibiotic resistant genes in *E. faecalis* isolates

Sample #	Source	Virulence traits[1]					Antibiotic resistance traits[1]							
		Cyl-A	Cyl-B	Cyl-M	gel-E	esp	pbp-5	gyrA	aac(6')-aph(2')	vanA	vanB	tetS	tetL	tetM
1	Endocervix	-[2]	-	-	-	+[3]	-	-	-	-	-	-	-	-
2	Fallopian tube (L)	-	-	-	-	-	-	-	-	-	-	-	-	-
3	Fallopian tube (L)	-	-	-	-	-	-	-	-	-	-	-	+	+
4	Fallopian tube (L)	-	-	-	-	+	-	-	-	-	-	-	-	+
5	Fallopian tube (L)	-	-	-	-	+	-	-	-	-	-	-	+	-
6	Fallopian tube (R)	-	-	-	-	+	-	-	-	-	-	-	-	+
7	Fallopian tube (R)	-	-	-	-	-	-	-	-	-	-	-	+	-
8	Ovary	-	-	-	-	-	+	-	-	-	-	-	-	+
9	Tissue penis	-	-	-	+	-	-	-	-	-	-	-	-	-
10	Urine	-	-	-	-	+	-	-	-	-	-	-	-	-
11	Urine	-	+	+	+	+	-	-	+	-	-	-	-	+
12	Urine	-	-	-	+	+	-	-	-	-	-	-	-	+
13	Urine	-	+	+	-	+	-	-	-	-	-	-	-	+
14	Urine	-	+	+	+	-	-	-	-	-	-	-	-	-
15	Urine	-	+	+	-	+	-	-	-	-	-	-	-	+
16	Urine	-	-	-	-	+	-	-	-	-	-	-	-	+
17	Urine	-	+	+	-	+	-	-	-	-	-	-	-	+
18	Urine	+	+	+	-	+	-	-	-	-	-	-	-	+
19	Urine	-	-	-	-	+	-	-	-	-	-	-	-	+
20	Urine	-	+	+	+	+	-	+	-	-	-	-	+	+
21	Urine	-	+	+	-	+	-	-	-	-	-	-	-	+
22	Urine	-	+	+	+	-	-	+	+	-	-	-	+	+

		Virulence traits					Antibiotic resistance traits							
Sample #	Source	Cyl-A	Cyl-B	Cyl-M	gel-E	esp	pbp-5	gyrA	aac(6')-aph(2')	vanA	vanB	tetS	tetL	tetM
23	Urine	-	-	+	+	-	-	+	+	-	-	-	+	+
24	Urine	-	-	+	+	-	-	-	-	-	-	-	+	-
25	Urine	-	+	+	+	-	-	+	+	-	-	-	+	+
26	Urine	-	-	-	-	+	-	-	-	-	-	-	-	+
27	Urine	-	+	-	+	-	-	-	-	-	-	-	-	-
28	Urine	-	+	+	-	-	+	-	-	-	-	-	-	+
29	Urine	-	+	+	+	-	-	-	-	-	-	-	-	+
30	Urine	-	-	+	+	+	-	-	-	-	-	-	+	+
31	Urine	+	+	+	+	-	-	+	+	-	-	-	+	+
32	Urine	-	-	-	+	+	-	+	+	-	-	-	+	+
33	Urine	-	+	+	+	+	-	+	+	-	-	-	+	+
34	Urine	+	+	-	+	+	-	-	-	-	-	-	-	-
35	Urine	-	-	+	+	+	-	+	+	-	-	-	+	+
36	Urine	-	-	-	+	+	-	-	-	-	-	-	+	+
37	Urine	-	+	+	+	+	-	+	+	-	-	-	+	+
38	Urine	+	+	+	+	+	-	+	-	-	-	-	-	+
39	Urine	-	+	-	+	+	-	-	-	-	-	-	+	+
40	Urine	-	-	-	+	+	-	-	-	-	-	-	-	-
41	Urine	+	-	+	+	-	-	-	-	-	-	-	+	+
42	Urine	-	-	-	+	+	-	-	-	-	-	-	+	-
43	Urine	-	-	-	+	-	-	-	-	-	-	-	-	+

[1]Resistance to vancomycin (vanA, vanB), tetracycline (tet L, tet M, tet S), ciprofloxacin (gyrA), ampicillin (pbp5) and gentamicin (aac(6')-aph(2')) and genes encoding virulence factors cytolysins (cylA, cylB, cylM), gelatinase (gel E) and extracellular surface protein (esp); [2] indicates genes absent; [3] indicates genes present.

A distinct difference in the biofilm biochemical phenotype has recently been reported for clinical compared to commensal strains. Clinical *E. faecalis* biofilms demonstrated higher metabolic activity (higher glucose uptake and lower protein concentration) and lower biomass when compared to commensal strains, suggesting that the biofilm biochemical phenotype, and not just biofilm formation, may play a role in virulence [80].

The genome of pathogenic *E. faecalis* isolates is estimated to contain over 25% mobile or exogenously acquired DNA including a pathogenicity island, multiple conjugative and composite transposons, integrated plasmid gene phage regions and a high number of insertion sequences [21]. Several studies have documented genetic features of *E. faecalis* to identify genes specific to community and clinical lineages [62,81-84]. Virulence factors, antibiotic resistance genes, mobile genetic elements and multilocus sequence typing patterns are associated with the potential of *E. faecalis* to cause disease in humans. The question of whether distinct *E. faecalis* lineages can be distinguished between clinical and commensal has recently been addressed by Kim et al., [85]. They compared the genomes of 31 *E. faecalis* strains and found that *E. faecalis* strains contain more core genes but that no clade separation exists between clinical and commensal strains based on ortholog presence/absence between clinical and nonclinical/commensal strains. In general, the genomes of clinical and commensal *E. faecalis* strains lacked specific structural and functional features.

Successful gene transfer is dependent upon cell-cell signalling and the subsequent release of aggregation substance (AS) to initiate close contact between cells to allow gene transfer to occur. The key role of AS as a virulence factor is supported by data indicating that hospital acquired *E. faecalis* strains contain significantly higher concentrations of AS encoding genes than do community strains [86,87]. Pathogenicity islands (PAIs) are large, horizontally transmitted genetic elements that contribute to the rapid evolution of non-pathogenic organisms to pathogenic forms [88,89]. The PAI of *E. faecalis* is approximately 150kb encoding multiple factors, including those that contribute to its virulence. McBride et al., [90] examined patterns of PAI variation in *E. faecalis* strains from clinical and commensal sources. They found the PAIs highly variable in gene content, and the organization of the genes were found to correlate with genetic lineage in only a few, closely related isolates. In general, PAI regions in different strains did not correlate with the predicted relatedness of strains as determined by MLST analysis of housekeeping genes. This suggests that the PAIs of *E. faecalis* strains is evolving even more rapidly than the core genome, and that the PAIs accrue new traits and continue to evolve through deletion and addition of genes through rapidly occurring recombination events.

5. ANTIBIOTIC RESISTANCE

Infections with *E. faecalis* can be difficult to treat due to intrinsic resistance to several commonly used antibiotics and the ability of the species to acquire resistance to most currently available antibiotics, either by mutation or by receiving foreign genetic material through the transfer of plasmids and transposons [91-93]. Plasmid mobilization is a major mechanism for enterococcal antibiotic resistance arising from horizontal gene transfer [94]. The transfer of antibiotic resistance genes is largely by pheromone-responsive conjugative plasmids [95,96]. Sex-pheromone responsive plasmids also transfer virulence factors including bacteriocin and cytolysin genes and a large number of cryptic genes, with as yet unknown function [97]. Table 3 lists selected enterococcal plasmids, with plasmids pCF10, pAD1, pAM373, pMG2200 and pBEE99 described as pheromone-responsive plasmids, and pAMβ1, pAM830 and pRE25 described as broad host range plasmids, *i.e.* not only found in *E. faecalis*.

Table 3. Selected *E. faecalis* plasmids with sequenced genomes (adapted from [88])

Plasmid[1]	Size[1]	Antibiotic resistance[2]	Other traits[2]	Accession number[1]	Reference[2]
pCF10	67.7	Tet(M)	AS, UV	AY855841	[98]
pAD1	59.3		AS, Cyt, UV	**	[99]
pAM373	36.8		AS	AE002565	[100]
pMG2200	106.5	Van(B)	AS, Bac, UV	AB374546	[101]
pBEE99	80.6		AS, Bac, Bee, UV	GU046453	[102]
pAMβ1	27.8	MLS	-	GU128949	[103]
pAM830	45.0	MLS, Van(A)	-	**	[96,104]
pRE25	50.2	Cm, MLS	-	X92945	[105]

[1]From NCBI Entrez Genomes (http://www.ncbi.nlm.nih.gov/sites/genome); GenBank accession numbers are shown. Plasmid size is shown in kilobases (kb), and for pAD1 and pAMβ1, extracted from references.
**accession number for complete plasmid is not available; readers are directed to references for more information.
[2]Antibiotic resistance and other traits were extracted from references; -, no known antibiotic resistance traits encoded. Tet(M), tetracycline resistance; Cm, chloramphenicol resistance; MLS, macrolide, lincosamide, streptogramin B resistance; Van(B), vancomycin resistance; AS, aggregation substance; UV, ultraviolet resistance; Bee, biofilm enhancer in Enterococcus; Cyl, cytolysin; Bac, bacteriocin.

Antibiotic resistance plasmids existed long before the usage of antimicrobials and are present in all investigated reservoirs. Plasmid transmission to enterococci in food and the environment is taking place continuously and can be seen as contamination either of the plasmid-containing strains, or as a result of antimicrobial selection for their presence [106]. With the data available, it is clear that antibiotic resistance traits encoded by plasmids are present in *E. faecalis* isolated from humans, food and animals. The pheromone responsive plasmids encode a unique and very efficient conjugation system that allows the spread of these plasmids in *E. faecalis* isolates, and therefore these plasmids can be considered as highly effective vectors for antibiotic resistance dissemination [107]. As shown in Table 4, antibiotic-resistant *E. faecalis* have been isolated from foods, including cheese, meat and produce in numerous studies [108-116]. In particular, erythromycin, tetracycline, chloramphenicol, ciprofloxacin and rifampicin resistance are the most common antibiotic resistance determinants detected in *E. faecalis* isolated from various food sources. It is accepted that the wide use of tetracyclines in animal farming contributes to the wide-spread resistance to tetracyclines in enterococci isolated from food animals [107]. Spread of antibiotic resistance from animals to humans is also evident. A food route of transmission has been described for *E. faecalis* isolates from farm equipment, milk, cheese and human faecal samples [117]. Evidence of antibiotic resistant enterococci from animal sources that cause human infections are limited, but have been documented and clonal relationships between hospital- and swine-associated vancomycin-resistant enterococci has been described previously [118,119].

Table 4. Reported incidences of antibiotic resistances among *E. faecalis* strains isolated from foods (adapted from [107])

Resistance to antibiotic	Fermented sausage[1]	Fermented sorghum[2]	Cheese/other foods[3]	Produce[4]	Ready to eat foods[5]	Bryndza cheese[6]	Moroccan foods[7]	Pecorino Abruzzese cheese[8]	Retail beef/chicken/pork[9]	French cheeses[10]	Retail milk/beef/chicken[11]
	n = 5	na	n = 47	n = 38	n = 52	n = 49	n = 23	n = 28	n = 94/192/92	n = 79[12]	n = 415/154/107
Ampicillin	nd	na	2.1	nd	0	0	0	0	nd	nd	96.4/99.4/93.5
Penicillin	nd	na	12.8	0.6/0	0	nd	0	37.9	0	nd	nd
Erythromycin	93.3	na	63.8	8.1/4.5	26.4	22	22	48.3	2/47/8	70.9	nd
Tetracycline	86.7	na	44.7	89/39	41.5	nd	87	86.9	14/88/43	92.4	18.3/5.8/1.9
Chloramphenicol	93.3	na	31.9	3.1/0	24.5	nd	9	6.9	0/0/2	55.7	nd
Ciprofloxacin	46.7	na	27.7	0.6/0	0	2	61	0	0	nd	nd
Gentamicin	0	na	25.5	nd	nd	0	0	6.9	0/2/1	5.1	nd
Streptomycin	nd	na	46.8	nd	nd	0	17	0	3/37/11	nd	nd
Vancomycin	0	na	0	nd	0	0	9	0	0	nd	18.8/7.8/13.1
Rifampin	100	na	nd	nd	11.3	29	78	nd	nd	nd	nd

nd: not determined; na: not applicable as either no *E. faecalis* strains were investigated.

[1]Data were adapted from [127]. [2]Data were adapted from [128]. [3]Data were adapted from [129]. [4]Data were adapted from [112]. [5]Data were adapted from [130]. [6]Data were adapted from [131]. [7]Data were adapted from [116]. [8]Data were adapted from [132]. [9]Data were adapted from [133,134]. [10]Data were adapted from [135]. [11]Data were adapted from [136]. [12] Isolates were selected for antibiotic resistance, the incidence (%) relates to the collection of resistant strains isolated from cheeses.

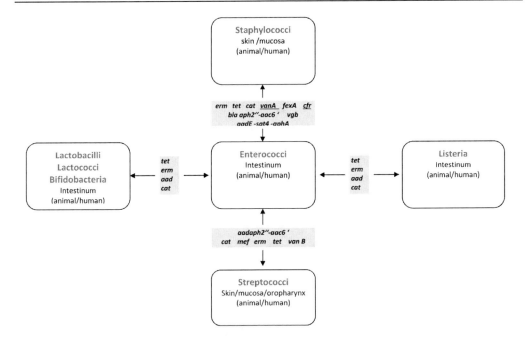

Figure 1. Resistance gene pool shared in Gram-positive bacteria. The second line describes the habitat of the corresponding bacterium. Resistances against: *aad*, strep-tomycin; *aac6-aph2*, gentamicin/tobramycin; *bla*, penicillins; *cat*, chloramphenicol; *cfr*, florfenicol/linezolid; *erm*, macrolide-lincosamide-streptogramin B (MLSB); *fexA*, florfenicol/chloramphenicol, *mef*, macrolides; *sat4*, nourseothricin; *tet*, tetracycline; *vanA*, vancomycin/teicoplanin; *vanB*, Vancomycin; *vgb*, streptogramin B. Some deter-minants represent several classes and types of resistance genes such as *erm* representing *erm(A)/(B)/(C)* or *tet* which stands for *tet(M)/(O)/(W)* (ribosomal protection) and *tet(K)/(L)* (efflux pumps) being summarized here for the sake of space. Underlined resistance genes are of special concern encoding resistance to important therapeutic substances or antibiotics of last resort (adapted from [99]).

Figure 1 indicates how antibiotic resistance genes can be shared between Gram-positive bacteria, in particular highlighting vancomycin and florfenicol/linezolid resistance traits that are of particular concern as these antibiotics are considered to be "last resort" antibiotics for treatment of human enterococcal infections.

In summary, there is evidence to suggest that food and animals can be considered sources of antibiotic resistant *E. faecalis* transmission to humans. Prudent use of antibiotics as therapeutic agents in animal husbandry and implementing alternatives to antibiotics have increasing importance in the prevention of more resistant strains and the transfer of these strains to humans via the food chain.

6. SHARED CLINICAL AND COMMENSAL CLONES IN *E. FAECALIS*

E. faecalis is a typical example of an opportunistic pathogen that has long been regarded as a relatively harmless commensal that colonizes the gastrointestinal tract of humans and animals. Although considered normal colonizers of the digestive tract, they have also been recognized as etiological agents of hospital-acquired infections in debilitated patients [120]. Multilocus sequence typing (MLST) is a sequence-based genotyping method that is performed on housekeeping genes. Using this method, two major clonal complexes (CCs) have been identified (CC2 and CC9), containing almost exclusively hospital-derived isolates, suggesting the adaptation of these clones to the hospital environment [121]. So far, the *E. faecalis* MLST scheme has proved its value in studying the hospital epidemiology of *E. faecalis* clones colonizing and causing infections in patients (14-17).

There are seven prevalent clinical *E. faecalis* clones based on MLST [121], ST6, ST9, ST16, ST21, ST28, ST40 and ST87. These STs account for 37% of the hospital-associated isolates (http:/efaecalis.mlst.net/), indicating that *E. faecalis* clones are more often shared between hospitalized patients and other reservoirs. Kuch et al., [122], performed molecular typing of 386 *E. faecalis* isolates from several European countries, and showed that STs belonging to six clonal complexes (CC2, CC16, CC21, CC30, CC40 and CC87) played a predominant role in the spread of antimicrobial resistance in hospitals. We performed an eBURST (Based Upon Related Sequence Types) analysis comparing all clinical and commensal *E. faecalis* strains listed in the MLST database as at March 2014. Figure 2 represents this comparison, and shows the overlapping ST's for both clinical and commensal *E. faecalis* STs. From this analysis, it is clearly evident that there is no clear distinction between clinical and commensal *E. faecalis* STs/clones, which has also been described by other authors. In a recent study by Buhnik-Rosenblau et al., [123], genomic typing (sequencing of SSR loci) of 106 *E. faecalis* strains isolated from diverse origins (animal feces, veterinary clinical isolates, human clinical isolates and milk-product strains), found one cluster representing the milk-product strains, and two clusters comprising a mixture of clinical and commensal isolates, which were dispersed along the phylogenetic dendrogram without forming any specific clusters.

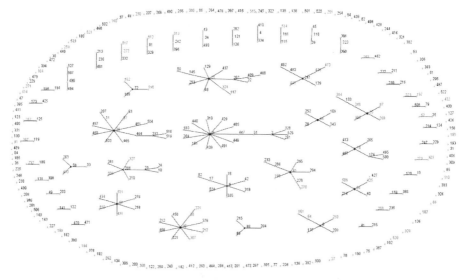

Figure 2. eBURST representation of both commensal and clinical sequence types. Black-clinical ; Green- commensal ; Pink- Present in both clinical and commensal.

Studies by Manson et al., [94] and Oliver & Green [124], reported that *E. faecalis* strains have a high abundance of transposable elements as well as an advanced pheromone system that enables it to not only spread plasmids, but also enables the easy transfer of large genomic regions (including virulence and resistance genes). These traits have increased the genome plasticity of *E. faecalis* strains, and it's adaptability to changing conditions in the hospital environment. *E. faecalis* multigenome analysis demonstrated the presence of an open "pan-genome", enabling efficient acquisition and integration of foreign DNA into the gene pool [125,126].

In the future, a systems biology approach that integrates transcriptome, proteome, metabolome analyses and functional studies will provide valuable insight into the population structure of *E. faecalis*. These investigations will enable us to understand the extent of lateral gene transfer and identify genetic lineages in *E. faecalis* isolates.

CONCLUSION

Based on current evidence, it appears that the transition of *E. faecalis* from a commensal to a pathogen is largely a consequence of its ability to transfer genomic elements to naive strains. An inherent difficulty in preventing and treating infections caused by pathogenic enterococci is the fact that avirulent commensal human isolates display significant heterogeneity in genotype. The endogenous *E. faecalis* population frequently harbours 'virulence' traits, which can be readily transferred to populations of enterococci that are perhaps transitional in the pathway from commensal to opportunistic pathogen. Whilst the preferred treatment option for community and hospital acquired bacterial infections continues to be antibiotics, it seems likely that this highly versatile opportunist will continue to actively acquire mechanisms of resistance to enhance its survival. A more prudent management strategy may in the future involve targeting genomic elements that favour mobilization and gene transfer.

REFERENCES

[1] Dunny GM (2013) Enterococcal sex pheromones: signaling, social behavior, and evolution. *Annu. Rev. Genet.* 47: 457-482.

[2] Christoffersen TE, Jensen H, Kleiveland CR, Dorum G, Jacobsen M, et al., (2012) In vitro comparison of commensal, probiotic and pathogenic strains of Enterococcus faecalis. *Br. J. Nut.r* 108: 2043-2053.

[3] Maddox SM, Coburn PS, Shankar N and Conway T (2012) Transcriptional regulator PerA influences biofilm-associated, platelet binding, and metabolic gene expression in Enterococcus faecalis. *PLoS One* 7: e34398.

[4] Hacker J and Kaper JB (2000) Pathogenicity islands and the evolution of microbes. *Annu. Rev. Microbiol.* 54: 641-679.

[5] Silhavy TJ, Kahne D and Walker S (2010) The bacterial cell envelope. *Cold Spring Harb. Perspect. Biol.* 2: a000414.

[6] Ryu YH, Baik JE, Yang JS, Kang SS, Im J, et al., (2009) Differential immunostimulatory effects of Gram-positive bacteria due to their lipoteichoic acids. *Int. Immunopharmacol.* 9: 127-133.

[7] Han SH, Kim JH, Seo HS, Martin MH, Chung GH, et al., (2006) Lipoteichoic acid-induced nitric oxide production depends on the activation of platelet-activating factor receptor and Jak2. *J. Immunol.* 176: 573-579.

[8] Buckley JM, Wang JH and Redmond HP (2006) Cellular reprogramming by gram-positive bacterial components: a review. *J. Leuko.c Biol.* 80: 731-741.

[9] Lotz S, Aga E, Wilde I, van Zandbergen G, Hartung T, et al., (2004) Highly purified lipoteichoic acid activates neutrophil granulocytes and delays their spontaneous apoptosis via CD14 and TLR2. *J. Leukoc. Biol.* 75: 467-477.

[10] Yang S, Tamai R, Akashi S, Takeuchi O, Akira S, et al., (2001) Synergistic effect of muramyldipeptide with lipopolysaccharide or lipoteichoic acid to induce inflammatory cytokines in human monocytic cells in culture. *Infect. Immun.* 69: 2045-2053.

[11] Ginsburg I (2002) Role of lipoteichoic acid in infection and inflammation. *Lancet Infect. Dis.* 2: 171-179.

[12] Periti P (2000) Current treatment of sepsis and endotoxaemia. *Expert Opin. Pharmacother* 1: 1203-1217.

[13] Beachey EH, Giampapa CS and Abraham SN (1988) Bacterial adherence. Adhesin receptor-mediated attachment of pathogenic bacteria to mucosal surfaces. *Am. Rev. Respir. Dis.* 138: S45-48.

[14] Ginsburg I and Kohen R (1995) Cell damage in inflammatory and infectious sites might involve a coordinated "cross-talk" among oxidants, microbial haemolysins and ampiphiles, cationic proteins, phospholipases, fatty acids, proteinases and cytokines (an overview). *Free Radic. Res.* 22: 489-517.

[15] Theilacker C, Kaczynski Z, Kropec A, Sava I, Ye L, et al., (2011) Serodiversity of opsonic antibodies against Enterococcus faecalis--glycans of the cell wall revisited. *PLoS One* 6: e17839.

[16] Geiss-Liebisch S, Rooijakkers SH, Beczala A, Sanchez-Carballo P, Kruszynska K, et al., (2012) Secondary cell wall polymers of Enterococcus faecalis are critical for resistance to complement activation via mannose-binding lectin. *J. Biol. Chem.* 287: 37769-37777.

[17] Kayaoglu G and Orstavik D (2004) Virulence factors of Enterococcus faecalis: relationship to endodontic disease. *Crit. Rev. Oral Biol. Med.* 15: 308-320.

[18] Lleo MM, Bonato B, Tafi MC, Signoretto C, Boaretti M, et al., (2001) Resuscitation rate in different enterococcal species in the viable but non-culturable state. *J. Appl. Microbiol.* 91: 1095-1102.

[19] Heim S, Lleo M, Bonato B, Guzman CA and Canepari P (2002) The viable but nonculturable state and starvation are different stress responses of Enterococcus faecalis, as determined by proteome analysis. *J. Bacteriol.* 184: 6739-6745.

[20] Tsikrikonis G, Maniatis AN, Labrou M, Ntokou E, Michail G, et al., (2012) Differences in biofilm formation and virulence factors between clinical and fecal enterococcal isolates of human and animal origin. *Microb. Pathog.* 52: 336-343.

[21] Upadhyaya GP, Lingadevaru UB and Lingegowda RK (2011) Comparative study among clinical and commensal isolates of Enterococcus faecalis for presence of esp gene and biofilm production. *J. Infect Dev. Ctries* 5: 365-369.

[22] Cosentino S, Podda GS, Corda A, Fadda ME, Deplano M, et al., (2010) Molecular detection of virulence factors and antibiotic resistance pattern in clinical Enterococcus faecalis strains in Sardinia. *J Prev Med Hyg* 51: 31-36.

[23] Hancock LE and Gilmore MS (2002) The capsular polysaccharide of Enterococcus faecalis and its relationship to other polysaccharides in the cell wall. *Proc. Natl. Acad. Sci. USA* 99: 1574-1579.

[24] Jankoska G, Trajkovska-Dokic E, Panovski N, Popovska-Jovanovska K and Petrovska M (2008) Virulence factors and antibiotic resistance in Enterococcus faecalis isolated from urine samples. *Prilozi* 29: 57-66.

[25] Simonsen GS, Smabrekke L, Monnet DL, Sorensen TL, Moller JK, et al., (2003) Prevalence of resistance to ampicillin, gentamicin and vancomycin in Enterococcus faecalis and Enterococcus faecium isolates from clinical specimens and use of antimicrobials in five Nordic hospitals. *J. Antimicrob. Chemother.* 51: 323-331.

[26] Leanti La Rosa S, Casey PG, Hill C, Diep DB, Nes IF, et al., (2013) In vivo assessment of growth and virulence gene expression during commensal and pathogenic lifestyles of luxABCDE-tagged Enterococcus faecalis strains in murine gastrointestinal and intravenous infection models. *Appl. Environ. Microbiol.* 79: 3986-3997.

[27] Vebo HC, Solheim M, Snipen L, Nes IF and Brede DA (2010) Comparative genomic analysis of pathogenic and probiotic Enterococcus faecalis isolates, and their transcriptional responses to growth in human urine. *PLoS One* 5: e12489.

[28] Vebo HC, Snipen L, Nes IF and Brede DA (2009) The transcriptome of the nosocomial pathogen Enterococcus faecalis V583 reveals adaptive responses to growth in blood. *PLoS One* 4: e7660.

[29] Shepard BD and Gilmore MS (1997) Identification of virulence genes in Enterococcus faecalis by differential display polymerase chain reaction. *Adv. Exp. Med. Biol.* 418: 777-779.

[30] Sharifi Y, Hasani A, Ghotaslou R, Naghili B, Aghazadeh M, et al., (2013) Virulence and antimicrobial resistance in enterococci isolated from urinary tract infections. *Adv Pharm Bull* 3: 197-201.

[31] Sharifi Y, Hasani A, Ghotaslou R, Varshochi M, Soroush MH, et al., (2012) Vancomycin-resistant enterococci among clinical isolates from north-west Iran: identification of therapeutic surrogates. *J. Med. Microbiol.* 61: 600-602.

[32] Kafil HS, Mobarez AM and Moghadam MF (2013) Adhesion and virulence factor properties of Enterococci isolated from clinical samples in Iran. *Indian J. Pathol. Microbiol.* 56: 238-242.

[33] van Merode AE, van der Mei HC, Busscher HJ, Waar K and Krom BP (2006) Enterococcus faecalis strains show culture heterogeneity in cell surface charge. *Microbiology* 152: 807-814.

[34] Waters CM and Bassler BL (2005) Quorum sensing: cell-to-cell communication in bacteria. *Annu. Rev. Cell Dev. Biol.* 21: 319-346.

[35] Carniol K and Gilmore MS (2004) Signal transduction, quorum-sensing, and extracellular protease activity in Enterococcus faecalis biofilm formation. *J. Bacteriol.* 186: 8161-8163.

[36] Podbielski A and Kreikemeyer B (2004) Cell density--dependent regulation: basic principles and effects on the virulence of Gram-positive cocci. *Int. J. Infect. Dis.* 8: 81-95.

[37] Arias CA and Murray BE (2012) The rise of the Enterococcus: beyond vancomycin resistance. *Nat Rev Microbiol* 10: 266-278.

[38] Waters CM, Hirt H, McCormick JK, Schlievert PM, Wells CL, et al., (2004) An amino-terminal domain of Enterococcus faecalis aggregation substance is required for aggregation, bacterial internalization by epithelial cells and binding to lipoteichoic acid. *Mol. Microbiol.* 52: 1159-1171.

[39] Theilacker C, Kaczynski Z, Kropec A, Fabretti F, Sange T, et al., (2006) Opsonic antibodies to Enterococcus faecalis strain 12030 are directed against lipoteichoic acid. *Infect. Immun.* 74: 5703-5712.

[40] Nallapareddy SR, Singh KV, Sillanpaa J, Garsin DA, Hook M, et al., (2006) Endocarditis and biofilm-associated pili of Enterococcus faecalis. *J. Clin. Invest.* 116: 2799-2807.

[41] Nallapareddy SR, Singh KV, Sillanpaa J, Zhao M and Murray BE (2011) Relative contributions of Ebp Pili and the collagen adhesin ace to host extracellular matrix protein adherence and experimental urinary tract infection by Enterococcus faecalis OG1RF. *Infect. Immun.* 79: 2901-2910.

[42] Sillanpaa J, Nallapareddy SR, Singh KV, Prakash VP, Fothergill T, et al., (2010) Characterization of the ebp(fm) pilus-encoding operon of Enterococcus faecium and its role in biofilm formation and virulence in a murine model of urinary tract infection. *Virulence* 1: 236-246.

[43] Brinster S, Posteraro B, Bierne H, Alberti A, Makhzami S, et al., (2007) Enterococcal leucine-rich repeat-containing protein involved in virulence and host inflammatory response. *Infect. Immun.* 75: 4463-4471.

[44] Xu Y, Jiang L, Murray BE and Weinstock GM (1997) Enterococcus faecalis antigens in human infections. *Infect. Immun.* 65: 4207-4215.

[45] Teng F, Jacques-Palaz KD, Weinstock GM and Murray BE (2002) Evidence that the enterococcal polysaccharide antigen gene (epa) cluster is widespread in Enterococcus faecalis and influences resistance to phagocytic killing of E. faecalis. *Infect. Immun.* 70: 2010-2015.

[46] Heikens E, Bonten MJ and Willems RJ (2007) Enterococcal surface protein Esp is important for biofilm formation of Enterococcus faecium E1162. *J. Bacteriol.* 189: 8233-8240.

[47] Fabretti F, Theilacker C, Baldassarri L, Kaczynski Z, Kropec A, et al., (2006) Alanine esters of enterococcal lipoteichoic acid play a role in biofilm formation and resistance to antimicrobial peptides. *Infect. Immun.* 74: 4164-4171.

[48] Theilacker C, Sanchez-Carballo P, Toma I, Fabretti F, Sava I, et al., (2009) Glycolipids are involved in biofilm accumulation and prolonged bacteraemia in Enterococcus faecalis. *Mol. Microbiol.* 71: 1055-1069.

[49] Beachey EH, Chiang TM, Ofek I and Kang AH (1977) Interaction of lipoteichoic acid of group A streptococci with human platelets. *Infect. Immun.* 16: 649-654.

[50] Courtney HS, Simpson WA and Beachey EH (1983) Binding of streptococcal lipoteichoic acid to fatty acid-binding sites on human plasma fibronectin. *J. Bacteriol.* 153: 763-770.

[51] Bhakdi S, Klonisch T, Nuber P and Fischer W (1991) Stimulation of monokine production by lipoteichoic acids. *Infect. Immun.* 59: 4614-4620.

[52] Saetre T, Kahler H, Foster SJ and Lyberg T (2001) Aminoethyl-isothiourea inhibits leukocyte production of reactive oxygen species and proinflammatory cytokines induced by streptococcal cell wall components in human whole blood. *Shock* 15: 455-460.

[53] Guiton PS, Hung CS, Hancock LE, Caparon MG and Hultgren SJ (2010) Enterococcal biofilm formation and virulence in an optimized murine model of foreign body-associated urinary tract infections. *Infect. Immun.* 78: 4166-4175.

[54] Rich RL, Kreikemeyer B, Owens RT, LaBrenz S, Narayana SV, et al., (1999) Ace is a collagen-binding MSCRAMM from Enterococcus faecalis. *J. Biol. Chem.* 274: 26939-26945.

[55] Shankar N, Lockatell CV, Baghdayan AS, Drachenberg C, Gilmore MS, et al., (2001) Role of Enterococcus faecalis surface protein Esp in the pathogenesis of ascending urinary tract infection. *Infect. Immun.* 69: 4366-4372.

[56] La Carbona S, Sauvageot N, Giard JC, Benachour A, Posteraro B, et al., (2007) Comparative study of the physiological roles of three peroxidases (NADH peroxidase, Alkyl hydroperoxide reductase and Thiol peroxidase) in oxidative stress response, survival inside macrophages and virulence of Enterococcus faecalis. *Mol. Microbiol.* 66: 1148-1163.

[57] Balla E, Dicks LM, Du Toit M, Van Der Merwe MJ and Holzapfel WH (2000) Characterization and cloning of the genes encoding enterocin 1071A and enterocin 1071B, two antimicrobial peptides produced by Enterococcus faecalis BFE 1071. *Appl. Environ. Microbiol.* 66: 1298-1304.

[58] Ike Y and Clewell DB (1984) Genetic analysis of the pAD1 pheromone response in Streptococcus faecalis, using transposon Tn917 as an insertional mutagen. *J. Bacteriol.* 158: 777-783.

[59] Ike Y, Hashimoto H and Clewell DB (1984) Hemolysin of Streptococcus faecalis subspecies zymogenes contributes to virulence in mice. *Infect. Immun.* 45: 528-530.

[60] Coburn PS, Pillar CM, Jett BD, Haas W and Gilmore MS (2004) Enterococcus faecalis senses target cells and in response expresses cytolysin. *Science* 306: 2270-2272.

[61] Key LL, Jr., Wolf WC, Gundberg CM and Ries WL (1994) Superoxide and bone resorption. *Bone* 15: 431-436.

[62] Qin X, Galloway-Pena JR, Sillanpaa J, Roh JH, Nallapareddy SR, et al., (2012) Complete genome sequence of Enterococcus faecium strain TX16 and comparative genomic analysis of Enterococcus faecium genomes. *BMC Microbiol.* 12: 135.

[63] Hill PA, Docherty AJ, Bottomley KM, O'Connell JP, Morphy JR, et al., (1995) Inhibition of bone resorption in vitro by selective inhibitors of gelatinase and collagenase. *Biochem. J.* 308 (Pt 1): 167-175.

[64] Hynes WL, Dixon AR, Walton SL and Aridgides LJ (2000) The extracellular hyaluronidase gene (hylA) of Streptococcus pyogenes. *FEMS Microbiol. Lett.* 184: 109-112.

[65] Weaver KE and Clewell DB (1989) Construction of Enterococcus faecalis pAD1 miniplasmids: identification of a minimal pheromone response regulatory region and evaluation of a novel pheromone-dependent growth inhibition. *Plasmid* 22: 106-119.

[66] Clewell DB (1993) Bacterial sex pheromone-induced plasmid transfer. *Cell* 73: 9-12.

[67] Pillai SK, Sakoulas G, Eliopoulos GM, Moellering RC, Jr., Murray BE, et al., (2004) Effects of glucose on fsr-mediated biofilm formation in Enterococcus faecalis. *J. Infect. Dis.* 190: 967-970.

[68] Baldassarri L, Bertuccini L, Ammendolia MG, Cocconcelli P, Arciola CR, et al., (2004) Receptor-mediated endocytosis of biofilm-forming Enterococcus faecalis by rat peritoneal macrophages. *Indian J. Med. Res.* 119 Suppl: 131-135.

[69] Barnes AM, Ballering KS, Leibman RS, Wells CL and Dunny GM (2012) Enterococcus faecalis produces abundant extracellular structures containing DNA in the absence of cell lysis during early biofilm formation. *MBio* 3: e00193-00112.

[70] Willems RJ and Bonten MJ (2007) Glycopeptide-resistant enterococci: deciphering virulence, resistance and epidemicity. *Curr. Opin. Infect. Dis.* 20: 384-390.

[71] Mohamed JA and Murray BE (2006) Influence of the fsr locus on biofilm formation by Enterococcus faecalis lacking gelE. *J. Med. Microbiol.* 55: 1747-1750.

[72] Mohamed JA, Teng F, Nallapareddy SR and Murray BE (2006) Pleiotrophic effects of 2 Enterococcus faecalis sagA-like genes, salA and salB, which encode proteins that are antigenic during human infection, on biofilm formation and binding to collagen type i and fibronectin. *J. Infect. Dis.* 193: 231-240.

[73] Davies DG, Parsek MR, Pearson JP, Iglewski BH, Costerton JW, et al., (1998) The involvement of cell-to-cell signals in the development of a bacterial biofilm. *Science* 280: 295-298.

[74] Lewis K (2001) Riddle of biofilm resistance. *Antimicrob Agents Chemother* 45: 999-1007.

[75] O'Toole G, Kaplan HB and Kolter R (2000) Biofilm formation as microbial development. *Annu. Rev. Microbiol.* 54: 49-79.

[76] Whitchurch CB, Tolker-Nielsen T, Ragas PC and Mattick JS (2002) Extracellular DNA required for bacterial biofilm formation. *Science* 295: 1487.

[77] Stoodley P, Sauer K, Davies DG and Costerton JW (2002) Biofilms as complex differentiated communities. *Annu. Rev. Microbiol.* 56: 187-209.

[78] Paganelli FL, Willems RJ and Leavis HL (2012) Optimizing future treatment of enterococcal infections: attacking the biofilm? *Trends Microbiol* 20: 40-49.

[79] Camargo IL, Zanella RC, Gilmore MS and Darini AL (2008) Virulence factors in vancomycin-resistant and vancomycin- susceptible Enterococcus faecalis from Brazil. *Braz J. Microbiol.* 39: 273-278.

[80] Meissner W, Jarzembowski TA, Rzyska H, Botelho C and Palubicka A (2013) Low metabolic activity of biofilm formed by isolated from healthy humans and wild mallards. *Ann. Microbiol.* 63: 1477-1482.

[81] Galloway-Pena J, Roh JH, Latorre M, Qin X and Murray BE (2012) Genomic and SNP analyses demonstrate a distant separation of the hospital and community-associated clades of Enterococcus faecium. *PLoS One* 7: e30187.

[82] Palmer KL, Godfrey P, Griggs A, Kos VN, Zucker J, et al., (2012) Comparative genomics of enterococci: variation in Enterococcus faecalis, clade structure in E. faecium, and defining characteristics of E. gallinarum and E. casseliflavus. *MBio* 3: e00318-00311.

[83] Solheim M, Brekke MC, Snipen LG, Willems RJ, Nes IF, et al., (2011) Comparative genomic analysis reveals significant enrichment of mobile genetic elements and genes

encoding surface structure-proteins in hospital-associated clonal complex 2 Enterococcus faecalis. *BMC Microbiol.* 11: 3.

[84] Lebreton F, van Schaik W, McGuire AM, Godfrey P, Griggs A, et al., (2013) Emergence of epidemic multidrug-resistant Enterococcus faecium from animal and commensal strains. MBio 4.

[85] Kim EB and Marco ML (2014) Nonclinical and clinical Enterococcus faecium strains, but not Enterococcus faecalis strains, have distinct structural and functional genomic features. *Appl. Environ. Microbiol.* 80: 154-165.

[86] Coque TM and Murray BE (1995) Identification of Enterococcus faecalis strains by DNA hybridization and pulsed-field gel electrophoresis. *J. Clin. Microbiol.* 33: 3368-3369.

[87] Berti M, Candiani G, Kaufhold A, Muscholl A and Wirth R (1998) Does aggregation substance of Enterococcus faecalis contribute to development of endocarditis? *Infection* 26: 48-53.

[88] Lindsay JA, Ruzin A, Ross HF, Kurepina N and Novick RP (1998) The gene for toxic shock toxin is carried by a family of mobile pathogenicity islands in Staphylococcus aureus. *Mol. Microbiol.* 29: 527-543.

[89] Carniel E, Guilvout I and Prentice M (1996) Characterization of a large chromosomal "high-pathogenicity island" in biotype 1B Yersinia enterocolitica. *J. Bacteriol.* 178: 6743-6751.

[90] McBride SM, Fischetti VA, Leblanc DJ, Moellering RC, Jr. and Gilmore MS (2007) Genetic diversity among Enterococcus faecalis. *PLoS One* 2: e582.

[91] Sood S, Malhotra M, Das BK and Kapil A (2008) Enterococcal infections & antimicrobial resistance. *The Indian journal of medical research* 128: 111-121.

[92] Hollenbeck BL and Rice LB (2012) Intrinsic and acquired resistance mechanisms in enterococcus. *Virulence* 3: 421-433.

[93] Hunt CP (1998) The emergence of enterococci as a cause of nosocomial infection. *British journal of biomedical science* 55: 149-156.

[94] Manson JM, Hancock LE and Gilmore MS (2010) Mechanism of chromosomal transfer of Enterococcus faecalis pathogenicity island, capsule, antimicrobial resistance, and other traits. *Proc. Natl. Acad. Sci. USA* 107: 12269-12274.

[95] Palmer KL, Carniol K, Manson JM, Heiman D, Shea T, et al., (2010) High-quality draft genome sequences of 28 Enterococcus sp. isolates. *J. Bacteriol.* 192: 2469-2470.

[96] Palmer KL, Kos VN and Gilmore MS (2010) Horizontal gene transfer and the genomics of enterococcal antibiotic resistance. *Curr. Opin. Microbiol.* 13: 632-639.

[97] Wirth R (1994) The sex pheromone system of Enterococcus faecalis. More than just a plasmid-collection mechanism? *Eur. J. Biochem.* 222: 235-246.

[98] Hirt H, Manias DA, Bryan EM, Klein JR, Marklund JK, et al., (2005) Characterization of the pheromone response of the Enterococcus faecalis conjugative plasmid pCF10: complete sequence and comparative analysis of the transcriptional and phenotypic responses of pCF10-containing cells to pheromone induction. *J. Bacteriol.* 187: 1044-1054.

[99] Francia MV, Haas W, Wirth R, Samberger E, Muscholl-Silberhorn A, et al., (2001) Completion of the nucleotide sequence of the Enterococcus faecalis conjugative virulence plasmid pAD1 and identification of a second transfer origin. *Plasmid* 46: 117-127.

[100] De Boever EH, Clewell DB and Fraser CM (2000) Enterococcus faecalis conjugative plasmid pAM373: complete nucleotide sequence and genetic analyses of sex pheromone response. *Mol. Microbiol.* 37: 1327-1341.

[101] Zheng B, Tomita H, Inoue T and Ike Y (2009) Isolation of VanB-type Enterococcus faecalis strains from nosocomial infections: first report of the isolation and identification of the pheromone-responsive plasmids pMG2200, Encoding VanB-type vancomycin resistance and a Bac41-type bacteriocin, and pMG2201, encoding erythromycin resistance and cytolysin (Hly/Bac). *Antimicrob Agents Chemother* 53: 735-747.

[102] Coburn PS, Baghdayan AS, Craig N, Burroughs A, Tendolkar P, et al., (2010) A novel conjugative plasmid from Enterococcus faecalis E99 enhances resistance to ultraviolet radiation. *Plasmid* 64: 18-25.

[103] Clewell DB, Yagi Y, Dunny GM and Schultz SK (1974) Characterization of three plasmid deoxyribonucleic acid molecules in a strain of Streptococcus faecalis: identification of a plasmid determining erythromycin resistance. *J. Bacteriol.* 117: 283-289.

[104] Flannagan SE, Chow JW, Donabedian SM, Brown WJ, Perri MB, et al., (2003) Plasmid content of a vancomycin-resistant Enterococcus faecalis isolate from a patient also colonized by Staphylococcus aureus with a VanA phenotype. *Antimicrob. Agents Chemother* 47: 3954-3959.

[105] Schwarz FV, Perreten V and Teuber M (2001) Sequence of the 50-kb conjugative multiresistance plasmid pRE25 from Enterococcus faecalis RE25. *Plasmid* 46: 170-187.

[106] Macedo AS, Freitas AR, Abreu C, Machado E, Peixe L, et al., (2011) Characterization of antibiotic resistant enterococci isolated from untreated waters for human consumption in Portugal. *International Journal of Food Microbiology* 145: 315-319.

[107] Werner G, Coque TM, Franz CM, Grohmann E, Hegstad K, et al., (2013) Antibiotic resistant enterococci-tales of a drug resistance gene trafficker. *Int. J. Med. Microbiol.* 303: 360-379.

[108] Busani L, Del Grosso M, Paladini C, Graziani C, Pantosti A, et al., (2004) Antimicrobial susceptibility of vancomycin-susceptible and -resistant enterococci isolated in Italy from raw meat products, farm animals, and human infections. *Int. J. Food Microbiol.* 97: 17-22.

[109] Christensen EA, Joho K and Matthews KR (2008) Streptogramin resistance patterns and virulence determinants in vancomycin-susceptible enterococci isolated from multi-component deli salads. *Journal of Applied Microbiology* 104: 1260-1265.

[110] Hayes JR, English LL, Carter PJ, Proescholdt T, Lee KY, et al., (2003) Prevalence and Antimicrobial Resistance of Enterococcus Species Isolated from Retail Meats. *Applied and Environmental Microbiology* 69: 7153-7160.

[111] Hummel A, Holzapfel WH and Franz CMAP (2007) Characterisation and transfer of antibiotic resistance genes from enterococci isolated from food. *Systematic and Applied Microbiology* 30: 1-7.

[112] Johnston LM and Jaykus L-A (2004) Antimicrobial Resistance of Enterococcus Species Isolated from Produce. *Applied and Environmental Microbiology* 70: 3133-3137.

[113] Miranda JM, Guarddon M, Mondragón A, Vázquez BI, Fente CA, et al., (2007) Antimicrobial resistance in Enterococcus spp. strains isolated from organic chicken,

conventional chicken, and turkey meat: A comparative survey. *Journal of Food Protection* 70: 1021-1024.

[114] Novais C, Coque TM, Boerlin P, Herrero I, Moreno MA, et al., (2005) Vancomycin-resistant Enterococcus faecium clone in swine, Europe [11]. *Emerging Infectious Diseases* 11: 1985-1987.

[115] Pérez-Pulido R, Abriouel H, Ben Omar N, Lucas R, Martínez-Cañamero M, et al., (2006) Safety and potential risks of enterococci isolated from traditional fermented capers. *Food and Chemical Toxicology* 44: 2070-2077.

[116] Valenzuela AS, Benomar N, Abriouel H, Cañamero MM and Gálvez A (2010) Isolation and identification of Enterococcus faecium from seafoods: Antimicrobial resistance and production of bacteriocin-like substances. *Food Microbiology* 27: 955-961.

[117] Gelsomino R, Vancanneyt M, Cogan TM and Swings J (2003) Effect of Raw-Milk Cheese Consumption on the Enterococcal Flora of Human Feces. *Applied and Environmental Microbiology* 69: 312-319.

[118] Larsen J, Schønheyder HC, Singh KV, Lester CH, Olsen SS, et al., (2011) Porcine and human community reservoirs of Enterococcus faecalis, Denmark. *Emerging Infectious Diseases* 17: 2395-2397.

[119] Freitas AR, Coque TM, Novais C, Hammerum AM, Lester CH, et al., (2011) Human and swine hosts share vancomycin-resistant Enterococcus faecium CC17 and CC5 and Enterococcus faecalis CC2 clonal clusters harboring Tn1546 on indistinguishable plasmids. *Journal of Clinical Microbiology* 49: 925-931.

[120] Murray BE (1990) The life and times of the Enterococcus. *Clin Microbiol Rev* 3: 46-65.

[121] Ruiz-Garbajosa P, Bonten MJ, Robinson DA, Top J, Nallapareddy SR, et al., (2006) Multilocus sequence typing scheme for Enterococcus faecalis reveals hospital-adapted genetic complexes in a background of high rates of recombination. *J. Clin. Microbiol.* 44: 2220-2228.

[122] Kuch A, Willems RJ, Werner G, Coque TM, Hammerum AM, et al., (2012) Insight into antimicrobial susceptibility and population structure of contemporary human Enterococcus faecalis isolates from Europe. *J. Antimicrob. Chemother.* 67: 551-558.

[123] Buhnik-Rosenblau K, Matsko-Efimov V, Danin-Poleg Y, Franz CM, Klein G, et al., (2013) Biodiversity of Enterococcus faecalis based on genomic typing. *Int. J. Food Microbiol.* 165: 27-34.

[124] Oliver KR and Greene WK (2009) Transposable elements: powerful facilitators of evolution. *Bioessays* 31: 703-714.

[125] Nelson KE, Weinstock GM, Highlander SK, Worley KC, Creasy HH, et al., (2010) A catalog of reference genomes from the human microbiome. *Science* 328: 994-999.

[126] van Schaik W, Top J, Riley DR, Boekhorst J, Vrijenhoek JE, et al., (2010) Pyrosequencing-based comparative genome analysis of the nosocomial pathogen Enterococcus faecium and identification of a large transferable pathogenicity island. *BMC Genomics* 11: 239.

[127] Martin B, Garriga M, Hugas M and Aymerich T (2005) Genetic diversity and safety aspects of enterococci from slightly fermented sausages. *J. Appl. Microbiol.* 98: 1177-1190.

[128] Yousif NM, Dawyndt P, Abriouel H, Wijaya A, Schillinger U, et al., (2005) Molecular characterization, technological properties and safety aspects of enterococci from 'Hussuwa', an African fermented sorghum product. *J. Appl. Microbiol.* 98: 216-228.

[129] Franz CM, Muscholl-Silberhorn AB, Yousif NM, Vancanneyt M, Swings J, et al., (2001) Incidence of virulence factors and antibiotic resistance among Enterococci isolated from food. *Appl. Environ. Microbiol.* 67: 4385-4389.

[130] Baumgartner A, Kueffer M and Rohner P (2001) Occurence and antibiotic resistance of enterococci in various ready-to-eat foods. *Archiv. Lebensmittelhyg* 52: 1-24.

[131] Belicova A, Krzkova L, Krajcovic J, Jurkovic D, Sojka M, et al., (2007) Antimicrobial susceptibility of Enterococcus species isolated from Slovak Bryndza cheese. *Folia Microbiol.* (Praha) 52: 115-119.

[132] Serio A, Paparella A, Chaves-Lopez C, Corsetti A and Suzzi G (2007) Enterococcus populations in Pecorino Abruzzese cheese: biodiversity and safety aspects. *J. Food Prot.* 70: 1561-1568.

[133] Aslam M, Diarra MS, Checkley S, Bohaychuk V and Masson L (2012) Characterization of antimicrobial resistance and virulence genes in Enterococcus spp. isolated from retail meats in Alberta, Canada. *Int. J. Food Microbiol.* 156: 222-230.

[134] Aslam M, Diarra MS and Masson L (2012) Characterization of antimicrobial resistance and virulence genotypes of Enterococcus faecalis recovered from a pork processing plant. *J. Food Prot.* 75: 1486-1491.

[135] Jamet E, Akary E, Poisson MA, Chamba JF, Bertrand X, et al., (2012) Prevalence and characterization of antibiotic resistant Enterococcus faecalis in French cheeses. *Food Microbiol.* 31: 191-198.

[136] Chingwaru W, Mpuchane SF and Gashe BA (2003) Enterococcus faecalis and Enterococcus faecium isolates from milk, beef, and chicken and their antibiotic resistance. *J. Food Prot.* 66: 931-936.

In: *Enterococcus faecalis*
Editor: Henry L. Mack

ISBN: 978-1-63321-049-3
© 2014 Nova Science Publishers, Inc.

Chapter 5

HIGH-LEVEL GENTAMICIN RESISTANCE IN *ENTEROCOCCUS FAECALIS*: MOLECULAR CHARACTERISTICS AND RELEVANCE IN SEVERE INFECTIONS

Mónica Sparo[*], MD and G. Delpech
ESCS-Medicina, Universidad Nacional del Centro de la Provincia de Buenos Aires, Argentina

ABSTRACT

The genus *Enterococcus* belongs to the indigenous gastrointestinal microbiota of humans and animals, and is present in food of animal origin as well as in vegetables. Whenever these bacteria cause invasive infections, their eradication is difficult. This phenomenon is linked to the natural and acquired antimicrobial resistance of enterococci as well as the existence of cell components that can behave as virulence factors. Cytolysin production contributes with the severity of infectious diseases in animal models and in humans. It has been proven that 60% of *Enterococcus faecalis* strains isolated from different infections sites are cytolysin producers. Their presence is associated with a five-fold increase of death risk in bacteremic patients. Clinical and microbiological resolution of severe infections can be affected when they are caused by cytolysin-producers *E. faecalis* strains that display high-level gentamicin resistance (HLGR). Enterococci carry a wide variety of mobile genetic elements and they are regarded as a reservoir of acquired antimicrobial resistance genes for Gram-positive bacteria. Multiple antimicrobial resistance is common among enterococci and constitutes a relevant Public Health issue. In 1979, HLGR (MIC ≥ 500 µg/mL) in enterococci was reported for the first time. HLGR represents a significant therapeutic problem for human medicine, especially for patients with invasive infectious diseases that require bactericidal efficacy such as meningitis, endocarditis and osteomyelitis. The most frequent HLGR gene among enterococci is *aac (6´)-ie-aph (2´´)-Ia* that encodes AAC(6´)-APH(2´´), an enzyme with acetyl transferase and phospho transferase activities. This bifunctional gene confers resistance to aminoglycosides available for therapeutic use

---

[*] E-mail:msparo@vet.unicen.edu.ar (Dr. Mónica Sparo).

with the exception of streptomycin. As a consequence, the synergistic role (bactericidal effect) of aminoglycosides with cell wall-active agents such as ampicillin or vancomycin is precluded. Along the last decade, *E. faecalis*has emerged as a relevant health-care associated pathogen. An identical mechanism for HLGR has been reported for human, animal and food enterococcal strains. In addition, exchange of these resistance genes through horizontal transfer is feasible. Risk factors for the acquisition of infection with high-level gentamicin resistant enterococci have been identified: previous long-term antimicrobial treatment, number of prescribed antimicrobials, previous surgeries, perioperative antimicrobial prophylaxis, hospitalization term/ antimicrobial treatment, urinary catheterization and renal failure. Infections caused by high-level gentamicin resistant *E. faecalis* constitute a severe risk for patients with invasive conditions and long-term hospitalization. Clonal expansion and emergence of unique bacterial strains contribute to the significant enhancement of infectious diseases caused by high-level gentamicin resistant *E. faecalis*.

INTRODUCTION

Enterococci are Gram-positive, facultative anaerobic cocci, with growth temperatures ranging from 10°C to 45°C, and an optimal growth rate at 35°C. Almost all enterococcal strains are homofermentatives, producing lactic acid from fermentation of glucose, without gas release. They are characterized by their ability to grow in broth with 6.5% NaCl as well as to hydrolyzeaesculin in the presence of 40% bile. The term "enterococci" was coined to stand out their intestinal origin, but the genus *Enterococcus* was not created until 1984. Most of these strains synthesized a cell-wall associated antigen (Lancefield´s group D antigen), although its detection is quite difficult and depends on the extraction procedure as well as on the quality of the antiserum used. For this genus, G + C content in the DNA varies between 37-45 mol%. Phylogenetic studies based on comparative analysis of 1,400 bases from 16S rRNA sequence revealed that the genus *Enterococcus* was most closely to *Vagococcus*, *Tetragenococcus* and *Carnobacterium* than with the phenotypically related genuses *Streptococcus* and *Lactococcus* (Murray, 1990).

Enterococci are capable of grow and survive under harsh environmental conditions; these bacteria have been found on soil, plants, animals, birds and insects. Enterococci are part of the indigenous gastro-intestinal microbiota of humans and animals. The microbiological and ecological factors that contribute with intestinal colonization are not known. Up to 10^8 CFU/g of enterococci have been found in human feces. In addition, strains from this genus have been isolated from fermented and dairy products. Moreover, someenterococcal strains have been regarded as probiotics (Castro et al., 2008; Castro et al., 2007; Sparo et al., 2012a; Sparo et al., 2008; Sparo & Mallo, 2001).

At present, 48 species have been proposed as members of the genus *Enterococcus* according to a compilation by Euzéby (2013). However, most of the human enterococcal infections are caused by *Enterococcus faecalis* and *Enterococcus faecium*, in a ratio 9:1.

For enterococci, there is scarce documentation about their colonization-associated factors, intestinal translocation, tissue adhesion, evasion of immune response and modulation of host`s inflammatory response. One of the most studied enterococcal virulence factors is cytolysin, which is widely produced by *E. faecalis* and it is linked to the rupture of cell membranes. Cytolysin production contributes with the severity of infectious diseases in animal models as well as in man. Previous studies have shown that cytolysin is produced by 60% of *E. faecalis* strains isolated from different infection sites. In addition, this virulence factor is associated with a five-fold increase of death among bacteremic patients(Huycke et al., 1991; Huycke et al., 1995; Jett et al., 1992).

Enterococci express natural (intrinsic) resistance to antimicrobial agents such as cephalosporins, clindamycin and trimethoprim-sulfametoxazole. In addition, enterococci have intrinsic low-level resistance to aminoglycosides because of impaired uptake. Minimal inhibitory concentrations (MICs) values range between 4 µg/mL and 256 µg/mL, and specifically MIC of gentamicin varies from 6µg/mL to 48 µg/mL (Chow, 2000).

Enterococci harbor a wide variety of mobile genetic elements and they are considered as a reservoir for acquired antimicrobial resistance genes among Gram-positive bacteria (Courvalin, 2006; Hegstad et al., 2010).

Acquired resistance to β-lactams is due to modification of penicillin-binding proteins (PBPs), especially PBP-5, as well as to β-lactamase production. In *E. faecalis*, the former resistance mechanism is very uncommon.

Enterococci show intrinsically low level of resistance to aminoglycosides. Nevertheless, they can express acquired high-level resistance to these antimicrobials which is due to the production of aminoglycoside-modifying enzymes.

Aminoglycosides bind to the 30S ribosomal subunit, which plays a crucial role in providing high-fidelity translation of genetic material, rendering the ribosome unavailable for translation and thereby resulting in cell death (Kotra et al., 2000; Recht & Puglisi, 2001).

Penicillins and glycopeptides are considered as bacteriostatic antimicrobials against enterococci (Landman & Quale, 1997). Synergism between these drugs and aminoglycosides presumably arises as the result of enhanced intracellular uptake of aminoglycosides caused by the increased permeability of bacteria after incubation with cell wall synthesis inhibitors such as β-lactams and glycopeptides (Moellering & Weinberg, 1971).

The main mechanism of high level aminoglycoside resistance in Gram-negative and Gram-positive bacteria is the enzymatic modification of drugs. Three families of enzymes have been avowed: aminoglycoside phosphotransferases (APHs, produce high-level resistance), aminoglycoside acetyltransferases (AACs) and aminoglycoside nucleotidyl transferases (ANTs).

The bifunctional enzyme AAC(6')-Ie-APH(2")-Ia found in enterococcal, streptococcal and staphylococcal isolates renders them high level resistance (MIC >2,000 µg/mL) to all clinically available aminoglycosides (gentamicin, amikacin, tobramicin, netilmicin and kanamicin), except for streptomycin and to some extent, arbekacin (Leclerq et al., 1992). Genes encoding aminoglycoside-modifying enzymes are often located on plasmids, which allow cell-to-cell dissemination of the aminoglycoside resistance trait. The *aac(6')-Ie-aph(2")-Ia*gene is generally flanked by inverted repeats of IS*256*, making up composit transposons such as Tn*5281* in *E. faecalis* which promotes a rapid dissemination at a molecular level (Hodel-Christian & Murray, 1991; Mingeot-Leclercq et al., 1999).

High-level gentamicin resistance (HLGR, MIC ≥ 500 µg/mL) in enterococci was firstly described in France (1979) and afterwards in the United States of America, at 1983 (Horodniceanu*et al.*, 1979; Mederski-Samoraj & Murray, 1983). Among *E. faecalis* and other enterococcal species with HLGR the most frequent gene is *aac (6')-Ie-aph (2'')-Ia*. This gene encodes the enzyme AAC(6')-APH(2"), which expresses acetyltransferase and phosphotransferase activities (Chow, 2000). The binfuctional gene is responsible for resistance to the clinically available aminoglycosides and precludes the synergism (bactericidal effect) between aminoglycosides and cell-wall active agents, such as ampicillin or vancomycin (Chow et al., 1998; Chow et al., 2001;Kobayashi et al., 2001; Leclerqet al., 1992). This genetic determinant has also been detected in other Gram-positive bacteria such as *Staphylococcus aureus, Streptococcus agalactiae* (Lancefield's Group B), *Streptococcus mitis* and Lancefield's Group G *Streptococcus* (Buu-Hoï et al., 1990; Galimand et al., 1999).

Other monofunctional genes that encode aminoglycoside-modyfying enzymes have been characterized. Among class APH (2˝)-subclass I phosphotransferases, chromosomal (e.g. *aph(2˝)-Ib* y *aph(2˝)-Id*) and plasmidic (e. g. *aph(2˝)-Ic*) genes have been described. These genes were originally found on *Enterococcus* species different than *E. faecalis*and encode enzymes that confer resistance to gentamicin and amikacin. The*aph(2˝)-Ic* gene is associated with gentamicin MIC`s from 128 to 512 μg/mL. Conversely, *aph(2")-Id* gene which was initially described in human *E. casseliflavus*, is linked to HLGR but not to amikacin resistance. This gene has been detected in several vancomycin-resistant *E. faecalis* isolates recovered from hospitalized patients (Ramírez & Tomalsky, 2010).

Other genes such as *aph(3´)-IIIa* and *ant(4´)-Ia* encode resistance to a wide number of aminoglycosides, excepting gentamicin (Chow et al., 1997). High-level streptomycin resistance (MIC ≥ 1,024 μg / mL) is chromosomally encoded and ribosomally mediated. This resistance is due to change on a protein (S12) located in the 30S ribosomal subunit or to the production of AAC (6 ´), an aminoglycoside adenyltranspherase (Eliopulos et al., 1984). Other types of high-level resistance to aminoglycosides found in *E. faecium* and *Escherichiacoli*, seem to be associated with *aac (6 ´)-Im*and *aph(2")-Ib* genes. The former resistance determinant has been described in vancomycin-resistant *E. faecium* strains isolated from hospitalized patients (Chow et al., 2001).

The resistance determinant *aac (6 ´) Ie-aph (2 ")* has been found in plasmids and it also was identified as part of a transposon, chromosomally or plasmidic located. This conjugative transposon was characterized as *Tn5281* by Hodel-Christian & Murray (1991).The composite transposon *Tn5281* (*IS256*-related), responsible for the expression of HLGR, is part of a conjugative plasmid (Hodel-Christian & Murray, 1991; Simjee et al., 2000). It is important to high-light that HLGR determinants have generally been found on plasmids in enterococci (Simjee et al., 2000) but is not known if there are particular plasmid types that confer gentamicin resistance. HLGR in enterococci is generally detected by assessing growth at high concentrations of gentamicin (120 μg) disk on Mueller Hinton agar (disc diffusion screening method). Interpretation is as follows: zone of inhibition = 6mm indicates high-level resistance to gentamicin; zone diameter ≥ 10mm indicates that the isolate is susceptible to high-level of gentamicin. Zone diameter of 7-9 mm is inconclusive and it is necessary to perform agar dilution or broth microdilution tests to confirm HLGR (MIC≥ 500 μg/mL). It is necessary the use of control strains such as *S. aureus* ATCC 25923 and *E. faecalis* ATCC 29212 (Chow, 2000; CLSI, 2014; EUCAST, 2014).

NON-HUMAN RESERVOIRS OF *E. FAECALIS* WITH HLGR

Since 1986, research regarding enterococcal acquired antimicrobial resistance has focused mainly on glycopeptide-resistant strains (vancomycin). In Europe, this phenomenon was linked with the use of sub-therapeutical doses of avoparcin as animal growth promoter (Murray, 1990). However, on recent years, emergence of high-level resistance to aminoglycosides in enterococci has been reported.

E. faecalis has shown to acquire antimicrobial resistance, HLGR in particular, but resistance to ampicillin and vancomycin is infrequent.

During 2000–2002, in Denmark, the proportion of high level gentamicin resistant *E. faecalis* isolates increased from 2% to 6% in the pig population. This event coincided with the emergence of high level gentamicin resistant*E. faecalis* isolates among patients with infective endocarditis in North Denmark Region (DANMAP, 2002). Afterwards, Larsen et al. (2010) demonstrated that all of these isolates (human and pig origin) belonged to the same clonal group, suggesting that pigs were a source of high-level gentamicin resistant *E. faecalis* infection.

Currently, in Argentina, pig farming is an important economic activity. Specifically, intensive breeding of pigs represents a large reservoir of antimicrobial resistant-bacteria. The use of antimicrobials in feed as growth promoters in a prophylactic and therapeutic fashion is a regular practice. Widespread selection of antimicrobial resistant bacteria in animals' gut facilitates spread of multi-resistant strains to humans through environment and food chain. Recently, Delpech et al. (2014) investigated the antimicrobial resistance with Public Health impact in *E. faecalis* isolated from pig feces of intensive breeding farms located at South-East region of Buenos Aires Province (Argentina). The *aac (6´)-Ie-aph (2´´)-Ia*gene conferring resistance to aminoglycosides was detected in *E. faecalis* isolates.

Choi & Woo performed a study with 101 food-borne *E. faecalis* isolates collected between 2003 and 2010; 11 high level gentamicin resistant enterococci (MIC >2,048 µg/mL) were found. All HLGR *E. faecalis* isolates encoded *aac(6')-Ie-aph(2")-Ia* and harbored at least 3 virulence traits among*asa1, esp, gelE, efaA, ace*, and *cylA* genes. Pulsed-field gel electrophoresis (PFGE) and multi-locus sequence typing (MLST) were carried out to characterize their molecular epidemiology. A total of 8 sequence types (STs), including 3 novel STs, were identified (ST35, ST82, ST116, ST202, ST300, ST403, ST407, and ST420). The STs of food-borne HLGR *E. faecalis* in this study have been confirmed as corresponding to clinical isolates in the MLST database (DB), except for ST300 and the new STs. Three out of 11 isolates belonged to CC116, including ST116, ST407, and ST420. This study provided evidence for the spread of HLGR *E. faecalis* with virulence factors to chicken sources in Korea. Authors concluded that chicken could be a potential transmission source of antimicrobial resistance and virulence factors (Choi & Woo, 2013).

In a collaborative study in six European countries, ca. 25.5% of HLGRin *E. faecalis* isolated from cheese was observed (Franz et al., 2001). More recently, in France, Jamet et al. (2012) studied cheeses made from unpasteurized milk and detected HLGR in 7% of the investigated *E. faecalis*.

In the South East of the Province of Buenos Aires (Argentina) one of the main economic activities are agriculture and livestock, where the manufacture of dairy products of different origin (cattle, sheep, goats) is quite relevant. Recently, in this region it was communicated *in vitro* resistance to antimicrobials for clinical use in *Enterococcus* spp. isolated from sheep cheese. MICs for gentamicin, streptomycin, vancomycin, teicoplanin, ampicillin, imipenem, linezolid and tigecycline were determined. In 4 isolates of *E. faecalis*, plasmidic HLGR (MIC$_{gen}$ 1024 µg/mL) was detected. *In vitro* horizontal transfer of HLGR determinants was observed by plasmid conjugation between 4/4 *E. faecalis* isolated from sheep cheese and the human recipient strain *E. faecalis* JH2-SS (Figure 1).

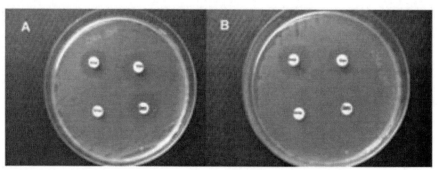

Source: Delpech et al. 2013.

Figure 1. *In vitro* transfer of plasmidic gentamicin resistance. A: *E. faecalis* JH2-SS, wild strain (not HLGR). B: *E. faecalis* JH2-SS, transconjugant (HLGR). HLGR: high-level gentamicin resistance. Amp: ampicillin; Van: vancomycin; Gen: gentamicin; Str: streptomycin.

Authors concluded that sheep cheese is a reservoir for resistant enterococci with potential spread to man through the food chain (Delpech et al., 2013).

A previous study that focused on investigating antimicrobial resistance profiles at phenotypic level of *E. faecalis* and *E. faecium* isolated from artisanal food of animal origin at the same region of Argentina was carried out. This study pointed out that artisanal products (cow cheese, goat cheese, artisanal salami and minced meat) could be considered as an environmental source of resistant enterococcal isolates to one or more antimicrobials such as β-lactams, glycopeptides and aminoglycosides, among other therapeutical agents (Delpech et al. 2012).

Sparo *et al.* (2013) investigated, along a one year period, the presence of cytolysin and HLGR in *E. faecalis* from human, animal and food origin. Clinical samples were obtained from patients with invasive infections in Hospital Ramón Santamarina fromTandilCity, Buenos Aires Province (Argentina). Chicken feces were taken from a farm at Tandil County. Minced meat was collected from butcher shops in Tandil City.Gene amplification was performed in order toidentify *E. faecalis* (*tuf* and *soda* genes) and to investigate *cylA* as well as HLGR coding genes. In addition, MICs for gentamicin and streptomycin were determined. *E. faecalis*isolates were recovered from chicken feces ($n = 41$), ground meat ($n = 38$), blood cultures ($n = 11$), sterile body fluids different from blood ($n = 9$) and retroperitoneal abscesses ($n = 2$). In chicken feces and minced meat *E. faecalis, cylA* was observed. In all the food isolates as well as in 7/9 animal isolates *cylA* and HLGR were associated. Among human samples, enterococci from blood cultures showed HLGR +*cylA*. In all enterococci with HLGR, *aac (6´) -Ie-aph (2´´)-Ia* gene was detected.

These findings matched with the MICs determined for gentamicin (> 500 μg/mL) and streptomycin (< 2,000 μg/mL). Detection of *aac (6´)-Ie-aph (2´´)-Ia* an d*cylA* in *E. faecalis* isolated from humans, food and animals proved its environmental spread (Figure 2).

A study carried out in Antofagasta, Chile, was in agreement with these findings. In *E. faecalis* isolated from clinical samples ($n = 52$) and chickens ($n = 28$), HLGR and virulence genes were detected (Silva et al. 2013).

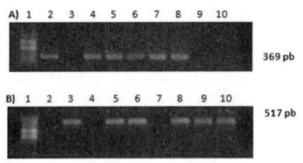

Source: Sparo et al. 2013.

Figure 2. Minced meat *E. faecalis* with high-level gentamicin resistance and cytolysin determinants. A) Genes amplification (multiplex PCR): *aac(6′)-Ie-aph(2″)-Ia*(369 pb), *aph(2″)-Ib* (867 pb), *aph(2″)-Ic* (444 pb), *aph(2″)-Id* (641 pb), *aph(3′)-IIIa*(523 pb) and *ant(4′)-Ia*(294 pb). Lane 1: molecular weight marker. Lane 2: *E. faecalis* NMH 524 (HLGR, 369 pb). Lane 3: *E. faecalis* ATCC 29212 (without HLGR). Lanes 4-8: minced meat isolates with HLGR (CPT13-CPT17). Lanes 9, 10: isolates without HLGR (CPT23, CPT25). B) *cylA* gene amplification (517 pb). Lane 1: molecular weight marker. Lane 2: *E. faecalis* CECT7121 without cytolysin. Lane 3: *E. faecalis* DS16 cytolysin producer. Lanes 4, 7: isolates (CPT10, CPT11) not producers of cytolysin. Lanes 5, 6, 8, 9, 10: isolates (CPT13-CPT17) with a 517 pb band.HLGR: High-level gentamicin resistance. pb: base pairs.

HLGR enterococci have also been found in aquatic environments such as well water and rivers. Presence of enterococcal strains with HLGR in secondary sewage effluent is of particular interest since enterococci have shown to be more resistant than *E. coli* to disinfection with chlorine (Rice et al., 1993).

The *aac(6')-Ie-aph(2")-Ia* resistance gene is highly prevalent in gentamicin resistant enterococcal isolates. As described above, several studies have demonstrated that *aac(6')-Ie-aph(2")-Ia* resistance gene is located on *Tn5281* transposable element. This fact suggests the possible contribution of this transposon on dissemination of resistance genes among enterococcal isolates (Zarrilli et al., 2005).

In patients with co-morbidity factors, a high-level of *E. faecalis* intestinal colonization can become a frequent precursor of human invasive infections through the translocation mechanism. This phenomenon is favored by the use of local or systemic broad-spectrum antimicrobials such as cephalosporins. The wide use of broad-spectrum antimicrobials contributes to the exertion of a significant genetic pressure upon intestinal microbiota. This turns out in an increasing emergency of multi-resistant bacteria. The human gastro-intestinal tract is a considerable reservoir of bacteria potentially capable of transfer resistance to conventional antimicrobials. Moreover, the fact that bacteria isolated from food of animal origin can behave as a resistance reservoir needs to be taken into consideration. *In vitro* studies performed to prove genetic exchange between enterococcal strains from humans and food of animal origin, are not conclusive. Therefore, *in vivo* models for assessing genetic transfer are needed. Research carried out in animal models with their own microbiota it will not be able to reproduce the conditions of the human intestine.

The use of human colon microbiota in germ-free mice is proposed as a model for reproducing the interaction between food strains and human gastro-intestinal microbiota (Hirayama, 1999).

Transfer of HLGR determinants from food to man was proven in an animal model by Sparo et al. (2012b). Immunocompetent BALB-C mice, colonized with human feces from an infant free of previous antimicrobial treatment, were used. This study supported the first evidence for the possibility of HLGR-gene transfer from food of animal origin enterococci to human enterococci. Therefore, a gene transfer model in non-sterile mice colonized with human gastro-intestinal microbiota was developed by these authors. Colonization of the gastro-intestinal tract with indigenous *E. faecalis* strains with HLGR implies the possibility of their translocation. Observed difficulties in antimicrobial therapy for invasive infections caused by high-level gentamicin resistant *E. faecalis* induce to assess their presence in food of animal origin.

HUMAN INFECTIOUS DISEASES DUE TO *E. FAECALIS* WITH HLGR

E. faecalis is found in the gastrointestinal and female urinary tracts as part of the indigenous host microbiota in healthy individuals where they cause infections. Current studies show that enterococci have emerged as a leading cause of nosocomial infections in hospitals and as community-acquired pathogens. These bacteria can cause serious invasive infections, including endocarditis, bacteremia, urinary tract and pelvic infections. Most of the enterococcal infections are due to *E. faecalis*. Clinical treatments for serious enterococcal infections require a combination of a cell wall active agent and an aminoglycoside, being gentamicin the most used. As mentioned above, in addition to the intrinsic resistance to several groups of drugs, enterococcal species can acquire high-level resistance to a variety of antimicrobials by horizontal transfer of mobile genetic determinants (Fisher & Phillips, 2009).

Recently, enterococci have been recognized as agents of severe infections with difficulties on their antimicrobial therapy, such as severe sepsis and infectious endocarditis in immune compromised patients.

Thepathogenic role of enterococci is generally underestimated; theyare capable of inducing severe bacteremia with high mortality, accompanied sometimes by endocarditis. Also, enterococcican produce intra-abdominal infections such as peritonitis and abscesses, and soft tissue infections such as those ones related with surgical wounds, ulcers and other types. Basis of treatment for serious infections is the combination of β-lactams with aminoglycosides, but resistance to any of them may be expressed, leading to treatment failure. It is important to have the results of these sensitivity tests as soon as possible to assess the need to include other antimicrobial options for the treatment such as glycopeptides or linezolid, depending on the possibility of resistance to the first (Arias et al., 2010; Gentile et al., 1995).

Enterococci, especially *E. faecalis*, are a significant cause of septicemia among newborns at neonatal intensive care units. Predisposing factors include: wide antimicrobials use, long-period staying for survival of low-weight newborns, extended use of parenteral nutrition and profuse use of vascular catheters (McNeeley et al., 1996).

Urinary tract is one of the most common targets of bacterial infections and it contributes to a significant amount of morbidity in the population. *E. coli* is the most commonly isolated pathogen, but *Enterococcus* spp.have emerged as important etiological agents showing resistance to a wide number of antimicrobials. *Enterococcus* species are more often associated with nosocomial urinary tract infections (Mims et al., 2006). In a Centers for Disease Control and Prevention's (CDC) survey of nosocomial infection, enterococci accounted for 13.9% of nosocomial urinary tract infection, after *E. coli* (Desai et al., 2001).

E. faecalis has an epidemic population structure dominated by a limited number of genetic lineages with an overrepresentation of clonal complexes (CC) CC2, CC9, CC10, CC16, CC21, CC30, CC40 and CC87. CC2, CC9 and CC87 constitute high risk CCs, as they are enriched in multi-drug resistant isolates causing infections in hospitalized patients (Ruiz-Garbajosa et al., 2006).

PFGE has been considered as the "gold standard" technique for the study of hospital outbreaks because of its high degree of strain differentiation. In Brazil, d' Azevedo et al. (2006) studied 455 clinical enterococcal isolates in five different hospitals and found agenetic diversity ranging from low (60.0%) to high similarity (95.0%).

However, MLST has emerged as an important tool for studying the long-term epidemiology and the population structure and patterns of evolutionary descent (Feil et al., 2004). Recently, in a study performed by Weng et al. (2013) in hospitals from Malaysia, MLST analysis of *E. faecalis* strains revealed the presence of ST6, which is associated with clonal-complex 2 (CC2). This finding deserves special attention since CC2 is commonly reported among nosocomial isolates and represents hospital-adapted complexes (Tomayko & Murray, 1995). Moreover, CC2 exhibits high-level resistance to aminoglycosides (Mato et al., 2009).

The seven most prevalent STs among clinical and outbreak-associated *E. faecalis* (ST6, ST9, ST16, ST21, ST28 ST40 and ST87), account for only 37% of the hospital associated isolates (Willems et al., 2011).

Some *E. faecalis* STs (ST16, ST21, ST28 and ST40) are also frequently found in the community, including farm animals and food isolates, showing low host specificity (Kuch et al., 2012).

Since the 1980s, an increase of HLGR in enterococci has been observed. Nevertheless, clinical significance of this resistance and its impact on therapeutic efficacy of severe infections are still being assessed. Among clinical studies focusing on HLGR in enterococci, stands out the one carried out by GarcíaVázquez et al. (2013). Authors detected bacteremia in2.2/1000 in-patients, caused mainly by *E. faecalis* (84%). In addition, in 83% of the patients an underlying condition was observed and 88% of the cases were hospital-acquired infections or health-care associated. Urinary tract infections were diagnosed among 20% of the patients meanwhile a higher frequency (47%) of primary bacteremia was observed. HLGR was detected (60%) but vancomycin or linezolid resistant strains were not reported. Multi-variate analysis pointed out that risk factors associated with HLGR were hospital-acquired infection (OR 6,083; CI95 % 1,428-25,915) and non-abdominal origin of infection (OR 6,006; IC95 % 1,398-25,805). Risk factors for mortality were initial clinical severity and previous empirical treatment (GarcíaVázquez et al., 2013).

During the period 2007-2009, in Japan, Araoka et al. (2011) studied epidemiological, microbiological and prognostic features of HLGR in enterococcal bacteremia, including severe cases of infectious endocarditis. A total of 155 episodes of enterococcal bacteremia were identified. Strains with HLGR accounted for 28% of the enterococcal strains, with predominance of HLGR in *E. faecalis* (32%) over *E. faecium* (24%). A mortality rate at 30 days of 31% was reported. A significant difference among mortality rates at 30 days, between enterococci with or without HLGR, was not observed. Authors considered that it is relevant to assess the presence or absence of HLGR expressing strains in all enterococcal infections with resistance to treatment, especially infectious endocarditis (Araoka et al., 2011).

Vigani et al. (2008) carried out a comparative study between 136 patients with bacteremia caused by HLGR enterococci and 79 patients with bacteremia caused by non-HLGR enterococci. Hematologic disease, neutropenia, *E. faecium* infection, nosocomial infection and monomicrobial bacteremia were more frequent in the group with HLGR than in enterococci without HLGR. Moreover, APACHE II scores were higher for the former group ($P<0.05$, in each case). Among independent risk factors for bacteremia caused by HLGR enterococci were included neutropenia, staying period in intensive care units previously to culture of enterococci with HLGR and use of third-generation cephalosporins. Univariate analysis showed that mortality at 14 or 30 days was higher for the group with HLGR than for the non-HLGR group (37% vs 15%, $P = 0.001$; 50% vs 22%, $P<0.001$). However, multivariate analysis determined that the presence of strains with HLGR it was not an independent risk factor for enterococcal bacteriemia mortality. Therefore, bacteremia caused by high level gentamicin resistant enterococci is associated with more conditions of severe co-morbidity and with a higher mortality than bacteremia caused by enterococci without HLGR. However, authors concluded that HLGR itself does not contribute in a significant way to mortality.

There is a significant variation of HLGR prevalence in enterococci from different geographic regions. In Tehran, a frequency of 65% was reported by Feizabadi et al. (2008). HLGR was reported in 46.15% of Italian enterococcal isolates (Zarrilli et al., 2005), in 45.5% of strains from Brazil (Vigani et al., 2008), 82.3% of enterococci in Michigan (Vakulenko et al., 2003), a lower frequency (37.64%) in Chicago, (Sahm & Gilmore, 1994) and in 46.06% of enterococci from South Africa (Keddy et al., 1996). The lowest rate (15.7%) of HLGR has been reported from Greece (Papaparaskevas et al., 2000).

However, it isimportant to note that the rate of HLGR in vancomycin-resistant enterococci (VRE) is higher than therate of HLGR for vancomycin-susceptible enterococci (VSE) strains. Mihajlović Ukropina et al. (2011) studied the frequency of antimicrobial resistance of enterococci isolated from blood cultures. HLGR was detected in VREstrains (87.6%) as well as in VSE strains (9.9%). Therefore, according to this study, HLGR in *E. faecium* appears to be higher than in *E. faecalis*.

In *E. faecalis*, HLGR and cytolysin production can condition the clinical and microbiological resolution of severe infectious diseases caused by these strains (Dupont et al., 2008).

Study of indigenousand opportunistic pathogenbacteria from gut microbiota suchas *E. faecalis* is relevant because when patients arrive at the hospital they may becolonized or carrying resistant bacteria to several antimicrobials, as well as they have increased risk of developing post-surgical infections from their own microbiota. These patients also bring to the hospital resistant bacteria that can be spread to other patients through the hands of healthcare staff, instrumental devices or other fomites. From the above mentioned it can be deduced that enterococci and specially, *E. faecalis*, are important opportunistic pathogens and cause of nosocomial infections, and hence the need to study the possible presence of this bacteria with antimicrobial resistance.

CONCLUSION

Risk factors for the acquisition of infections with high level gentamicin resistant enterococci are previous long-term antimicrobial treatment, number of prescribed antimicrobials, previous surgeries, perioperative antimicrobial prophylaxis, hospitalization term/antimicrobial treatment, urinary catheterization and renal failure. Prevalence of HLGR among *E. faecalis* has increased across the world and in general has been associated with the *aac(6')-Ie-aph(2")-Ia* gene. Several infectious diseases produced by *E. faecalis* are traditionally treated with a combination of cell wall active antimicrobials such as β-lactams or glycopeptides, and gentamicin. Therefore, increased rates of high level gentamicin resistant *E. faecalis* have led to the use of other and maybe less efficient drugs.

Infections caused by *E. faecalis* with HLGR constitute a severe risk for patients with invasive conditions and long-term hospitalization. Clonal expansion and emergence of unique bacterial strains contribute to the significant enhancement of infectious diseases caused by HLGR *E. faecalis* strains.

REFERENCES

Araoka, H; Kimura, M; Yoneyama, A. A surveillance of high-level gentamicin-resistant enterococcal bacteremia. *J. Infect. Chemother.* 2011; 17:433-434.

Arias, CA; Contreras, GA; Murray, BE. Management of multidrug-resistant enterococcal infections. *Clin. Microbiol. Infect.* 2010; 16: 555-62.

Buu-Hoï, A; Le Bouguenec, C; Horaud, T. High-level chromosomal gentamicin resistance in *Streptococcus agalactiae* (group B). *Antimicrob. Agents Chemother.* 1990; 34: 985–988.

Castro, M; Molina, M; Sparo, M;Manghi, M. Effects of *Enterococcus faecalis* CECT7121 on the specific immune response after DTPw vaccination. *Int. J. Prob. Preb.* 2008; 3: 25-30.

Castro, M; Sparo, M; Molin, M; Andino, J; Manghi, M. *Enterococcus faecalis* CECT7121 induces systemic immunomodulatory effects and protects from *Salmonella* infection. *Int. J. Prob. Preb.*2007; 2: 215-224.

Choi, JM; Woo, GJ. Molecular characterization of high-level gentamicin-resistant *Enterococcus faecalis* from chicken meat in Korea. *Int. J. Food Microbiol.*2013; 165:1-6.

Chow, JW, Aminoglycoside resistance in enterococci. *Clin. Infect. Dis.* 2000; 31: 586-589.

Chow, JW; Donabedian, SM; Clewell, SM; Sahm, DB; Zervos, MJ.*In vitro* susceptibility and molecular analysis of gentamicin-resistant enterococci. *Diagn. Microbiol. Infect. Dis.* 1998; 32: 141–146.

Chow, JW; Kak, V; You, I; Kao, SJ; Petrin, J; Clewell, DB; Lerner, SA; Miller, GH; Shaw, KJ. Aminoglycoside resistance genes *aph(2")-Ib*and*aac(6')-Im* detected together in strains of both *Escherichia coli and Enterococcus faecium*. *Antimicrob. Agents Chemother.* 2001;45: 2691-2694.

Chow, JW; Zervos, MJ; Lerner, SA. A novel gentamicin resistance gene in *Enterococcus*. *Antimicrob. Agents Chemother.* 1997; 41: 511–514.

CLSI. Clinical and Laboratory Standards Institute. Performance standards for antimicrobial susceptibility testing: 24th informational supplement. *CLSI Document M100-S24.* CLSI: Wayne, PA. 2014.

Courvalin, P. Vancomycin resistance in Gram-positive cocci. *Clin. Infect. Dis.* 2006; 42 (Suppl 1): S25–34.

DANMAP. Danish Integrated Antimicrobial Resistance Monitoring and Research Program (DANMAP 2002). Use of antimicrobial agents and occurrence of antimicrobial resistance in bacteria from food animals, *foods and humans in Denmark* [cited 2009 Sep 1]http://www.danmap. org/pdfFiles/Danmap_2002.pdf.

d'Azevedo, PA; Dias, CA; Teixeira, LM. Genetic diversity and antimicrobial resistance of enterococcal isolates from Southern region of Brazil. *Rev. Inst. Med. Trop. Sao Paulo.* 2006; 48: 11–16.

Delpech, G; Pourcel, G; Schell, C; De Luca, M; Basualdo, J; Bernstein, J; Grenovero, S; Sparo, M. Antimicrobial resistance profiles of *Enterococcus faecalis* and *Enterococcus faecium* isolated from artisanal food of animal origin in Argentina. *Foodborne Pathog. Dis.* 2012; 9: 939-944.

Delpech, G; Sparo, M; Pourcel, G; Schell, C; De Luca, MM; Basualdo, JA. *Enterococcus* spp. aislados de queso de oveja: resistencia a los antimicrobianos de utilización clínica. *Rev. Soc. Ven. Microbiol.* 2013; 33: 129-133.

Delpech, G; Sparo, MD; Pourcel, G;Ceci, M; Solana, V. *Enterococcus faecalis* isolated from pig feces: Antimicrobial resistance with Public Health Impact. *4th ASM Conference on Enterococci.* Cartagena, Colombia. March 5-7, 2014.

Desai, PJ; Pandit, D; Mathur, M, Gogate, A. Prevalence, identification and distribution of various species of enterococci isolated from clinical specimens with special reference to urinary tract infection in catheterized patients. *Indian J. Med. Microbiol.* 2001; 19: 132-155.

Dupont, H;Vael, C; Muller-Serieys, C; Chosidow, D; Mantz, J; Marmuse, JP; Andremont, A; Goossens, H; Desmonts, JM. Prospective evaluation of virulence factors of enterococci isolated from patients with peritonitis: impact on outcome. *Diagn. Microbiol. Infect. Dis.* 2008; 60: 247-253.

Eliopoulos, GM; Farber, BF; Murray, BE;Wennersten, C; Moellering, RC Jr. Ribosomal resistance of clinical enterococcal isolates to streptomycin. *Antimicrob. Agents Chemother.* 1984; 25: 398–399.

EUCAST. European Committee of Antimicrobial Susceptibility Testing. *EUCAST Breakpoint table v 4.0.European Society of Clinical Microbiology and Infectious Diseases.* 2014. Available at: http://www.eucast.org/clinical_breakpoints/.

Euzéby, JP. List of Prokaryotic names with standing in nomenclature; genus *Enterococcus*.www.bacterio.net/enterococcus.html. 2013 (Accessed September 24th, 2013).

Feil, EJ; Li, BC; Aanensen, DM; Hanage, WP; Spratt, BG. eBURST: inferring patterns of evolutionary descent among clusters of related bacterial genotypes from multilocus sequence typing data. *J. Bacteriol.* 2004; 186: 1518–1530.

Feizabadi, MM; Shokrzadeh, L; Sayady, S; Asadi, S. Transposon *Tn5281* is the main distributor of the aminoglycoside modifying enzyme gene among isolates of *Enterococcus faecalis* in Tehran hospitals. *Can. J. Microbiol.*2008; 54:887–890.

Fisher, K; Phillips, C. The ecology, epidemiology and virulence of *Enterococcus. Microbiology.*2009; 155: 1749–1757.

Franz, CMAP; Muscholl-Silberhorn, AB;Yousif, NMK; Vancanneyt, M; Swings, J;Holzapfel, WH. Incidence of virulence factors and antibiotic resistance among enterococci isolated from food. *Appl. Environ. Microbiol.* 2001; 67: 4385-4389.

Galimand, M; Lambert, T; Gerbaud, G; Courvalin, P. High-level aminoglycoside resistance in the beta-hemolytic group G *Streptococcus* isolate BM2721. *Antimicrob.Agents Chemother.*1999; 43: 3008–3010.

García-Vázquez, E; Albendín, H; Hernández-Torres, A; Canteras, M; Yagüe, G; Ruiz, J; Gómez, J. Study of a cohort of patients with *Enterococcus* spp. bacteraemia. Risk factors associated to high-level resistance to aminoglycosides. *Rev. Esp. Quimioter.* 2013; 26:203-213.

Gentile, JH; Sparo, MD; Pipo, VB; Gallo, AJ. Meningitis due to *Enterococcus faecalis*. *Medicina (B. Aires).* 1995; 55:435-437.

Hegstad, K; Mikalsen, T; Coque, TM; Werner, G; Sundsfjord, A. Mobile genetic elements and their contribution to the emergence of antimicrobial resistant *Enterococcus faecalis* and *Enterococcus faecium*. *Clin. Microbiol. Infect.* 2010; 16: 541-554.

Hirayama, K. Ex-germfree mice harboring intestinal microbiota derived from other animal species as an experimental model for ecology and metabolism of intestinal bacteria. *Exp. Anim.* 1999; 48: 219–227.

Hodel-Christian, SL & Murray, BE. Characterization of the gentamicin resistance transposon *Tn5281* from *Enterococcus faecalis* and comparison to staphylococcal transposons *Tn4001* and *Tn4031*. *Antimicrob. Agents Chemother.* 1991; 35: 1147–1152.

Horodniceanu, T; Bougueleret, L; El-Solh, N; Bieth, G; Delbos, F. High-level, plasmid-borne resistance to gentamicin in *Streptococcus faecalis* subsp. *zymogenes. Antimicrob. Agents Chemother.*1979; 16: 686–689.

Huycke, MM; Gilmore, MS. Frequency of aggregation substance and cytolysin genes among enterococcal endocarditis isolates. *Plasmid.*1995; 34:152–156.

Huycke, MM; Spiegel, CA; Gilmore, MS. Bacteremia caused by hemolytic, high-level gentamicin-resistant *Enterococcus faecalis*. *Antimicrob. Agents Chemother.* 1991; 35:1626–1634.

Jamet, E; Akary, E; Poisson, MA;Chamba, JF; Bertrand, X;Serror, P. Prevalence and characterization of antibiotic resistant *Enterococcus faecalis* in French cheeses. *Food Microbiol.* 2012; 31:191-198.

Jett, BD; Jensen, HG; Nordquist, RE; Gilmore, MS. Contribution of the pAD1-encoded cytolysin to the severity of experimental *Enterococcus faecalis* endophthalmitis. *Infect. Immun.* 1992; 60: 2445–2452.

Keddy, KH; Klugman, KP; Liebowitz, LD. Incidence of high-level gentamicin resistance in enterococci at Johannesburg Hospital. *S. Afr. Med. J.* 1996; 86: 1273-1276.

Kobayashi, N; Alam, M; Nishimoto, Y; Urasawa, S; Uehara, N; Watanabe, N. Distribution of aminoglycoside resistance genes in recent clinical isolates of *Enterococcus faecalis, Enterococcus faecium*and *Enterococcus avium. Epidemiol., Infect.*2001; 126: 197–204.

Kotra, LP; Haddad, J; Mobashery, S. Aminoglycosides: perspectives on mechanisms of action and resistance and strategies to counter resistance. *Antimicrob. Agents Chemother.*2000; 44: 3249-3256.

Kuch, A; Willems, RJ; Werner, G; Coque, TM; Hammerum, AM; Sundsfjord, A; Klare, I; Ruiz-Garbajosa, P; Simonsen, GS; van Luit-Asbroek, M; Hryniewicz, W; Sadowy, E. Insight into antimicrobial susceptibility and population structure of contemporary human

*Enterococcus faecalis*isolates from Europe. *J. Antimicrob. Chemother.* 2012; 67: 551-558.

Landman, D; Quale, JM. Management of infections due to resistant enterococci: a review of therapeutic options. *J. Antimicrob.Chemother.* 1997; 40: 161-170.

Larsen, J;Schønheyder, HC; Lester, CH; Olsen, SS; Porsbo, LJ; Garcia-Migura, L; Jensen, LJ; Bisgaard, M; Hammerum, AM. Porcine-origin gentamicin-resistant *Enterococcusfaecalis* in humans, Denmark. *Emerg. Infect. Dis.* 2010; 16: 682–684.

Leclercq, R; Dutka-Malen, S; Brisson-Noël, A; Molinas, C; Derlot, E; Arthur, M; Duval, J; Courvalin, P. Resistance of enterococci to aminoglycosides and glycopeptides. *Clin. Infect. Dis.* 1992; 15: 495–501.

Mato, R; Almeida, F; Pires, R; Rodrigues, P; Ferreira, T; Santos-Sanches, I. Assessment of high-level gentamicin and glycopeptide-resistant *Enterococcus faecalis* and *E. faecium* clonal structure in a Portuguese hospital over a 3-year period. *Eur. J. Clin. Microbiol. Infect. Dis.*2009; 28: 855–859.

McNeeley, DF; Saint-Louis, F; Noel, GJ. Neonatal enterococcal bacteremia: an increasingly frequent event with potentially untreatable pathogens. *Pediatr. Infect. Dis. J.* 1996; 15:800-805.

Mederski-Samoraj, BD; Murray, BE. High-level resistance to gentamicin in clinical isolates of enterococci. *J. Infect. Dis.* 1983; 147: 751–757.

MihajlovićUkropina, M; Jelesić, Z;Gusman, V; Milosavljević, B. Frequency of vancomycin-resistant *enterococci* isolated from blood cultures from 2008 to 2010. *Med. Pregl.*2011; 64:481-485.

Mims, C;Dockrell, HM; Goering, RV; et al. (eds.). Urinary tract infections. In: Medical Microbiology. 3rd ed. Philadelphia, PA: Elsevier Mosby. 2006. Pp. 241-244.

Mingeot-Leclercq, MP; Glupczynski, Y; Tulkens, PM. Aminoglycosides: activity and resistance. *Antimicrob. Agents Chemother.* 1999; 43: 727-737.

Moellering, RC, Jr; Weinberg, AN. Studies on antibiotic synergism against enterococci. II. Effect of various antibiotics on the uptake of 14 C-labeled streptomycin by enterococci. *J. Clin. Invest.* 1971; 50: 2580-2584.

Murray, BE. The life and times of the enterococcus. *Clin. Microbiol. Rev.* 1990; 3: 40-65.

Papaparaskevas, J; Vatopoulos, A; Tassios, PT; Avlami, A; Legakis, NJ; Kalapothaki, V. Diversity among high-level aminoglycoside-resistant enterococci. *J. Antimicrob. Chemother.* 2000; 45:277–283.

Ramirez, MS; Tolmasky, ME. Aminoglycoside modifying enzymes. *Drug Resist. Updat.* 2010; 13: 151-171.

Recht, M.I; Puglisi, JD. Aminoglycoside resistance with homogeneous and heterogeneous populations of antibiotic-resistant ribosomes. *Antimicrob. Agents Chemother.* 2001; 45: 2414-2419.

Rice, EW; Covert, TC; Wild, DK; Berman, D; Johnson, SA; Johnson. CH. Comparative resistance of *Escherichia coli* and enterococci tochlorination. *J. Environ. Sci. Health Part A.* 1993; 28:89–97.

Ruiz-Garbajosa, P; Bonten, MJ; Robinson, DA; Top, J; Nallapareddy, SR; Torres, C; Coque, TM; Cantón, R; Baquero, F; Murray, BE; del Campo, R; Willems, RJ. Multilocus sequence typing scheme for *Enterococcus faecalis* reveals hospital-adapted genetic complexes in a background of high rates of recombination. *J. Clin.Microbiol.*2006; 44: 2220-2228.

Sahm, DF; Gilmore, MS. Transferability and genetic relatedness of high-level gentamicin resistance among enterococci. *Antimicrob. Agents Chemother*. 1994;38:1194–1196.

Silva, J; Rodríguez, Y; Araya, J; Gahona, J; Valenzuela, N; Guerrero, K; Báez, J; Baquero, F; del Campo, R. Detection of virulence genes in aminoglycoside susceptible and resistant *Enterococcus faecalis. Rev. Chilena Infectol.*2013; 30: 17-22.

Simjee, S; Manzoor, SE; Fraise, AP; Gill, MJ. Nature of transposon-mediated high-level gentamicin resistance in *Enterococcus faecalis* isolated in the United Kingdom. *J. Antimicrob. Chemother*. 2000; 45: 565-575.

Sparo, M; Delpech, G; Pourcel, G; Schell, C; De Luca, MM; Bernstein, J; Grenóvero, S; Basualdo, JA. Citolisina y alto nivel de resistencia a gentamicina en *Enterococcusfaecalis* de distinto origen. *RAZ y EIE*. 2013; 8: 5-10.

Sparo, M; Mallo, R. Evaluación de la flora bacteriana de un ensilado natural de maíz. *Rev. Arg. Microbiol*. 2001; 33: 75-80.

Sparo, M; Nuñez, GG; Castro, M; Calcagno, ML; García Allende, MA; Ceci, M; Najle, R; Manghi, M.Characteristics of an environmental strain, *Enterococcus faecalis* CECT7121 and its effects as additive on craft dry-fermented sausages. *Food Microbiol*. 2008; 25: 607-615.

Sparo, M; Urbizu, L; Solana, MV; Pourcel, G; Delpech, G; Confalonieri, A; Ceci, M; Sánchez Bruni, SF. High level resistance to gentamicin: genetic transfer between *Enterococcus faecalis* isolated from food of animal origin and human microbiota. *Lett. Appl. Microbiol*. 2012b; 54: 119-125.

Sparo, MD; Corso, A; Gagetti, P; Delpech, G; Ceci, M; Confalonieri, A; Urbizu, L; Sánchez Bruni, SF. *Enterococcus faecalis* CECT7121: biopreservation of crafted goat cheese. *Int. J. Prob. Preb*. 2012a; 7: 145-152.

Tomayko, JF; Murray, BE. Analysis of *Enterococcus faecalis* isolates from intercontinental sources by multilocus enzyme electrophoresis and pulsed-field gel electrophoresis. *J. Clin. Microbiol.*1995; 33: 2903–2907.

Vakulenko, SB; Donabedian, SM; Voskresenskiy, AM; Zervos, MJ; Lerner, SA; Chow, JW. Multiplex PCR for detection of aminoglycoside resistance genes in enterococci. *Antimicrob. Agents Chemother*. 2003; 47:1423–1426.

Vigani, AG; Oliveira, AM; Bratfich, OJ; Stucchi, RS; Moretti, ML. Clinical, epidemiological, and microbiological characteristics of bacteremia caused by high-level gentamicin–resistant *Enterococcus faecalis. Braz. J. Med. Biol. Res*. 2008; 41:890-895.

Weng, PL; Ramli, R; Shamsudin, MN; Cheah, YK; Hamat, RA. High genetic diversity of *Enterococcus faecium* and *Enterococcus faecalis* clinical isolates by Pulsed-Field Gel Electrophoresis and Multilocus Sequence Typing from a hospital in Malaysia. *Biomed. Res. Int*. 2013; 2013: 938937.

Willems, RJ; Hanage, WP; Bessen, DE; Feil, EJ. Population biology of Gram-positive pathogens: high-risk clones for dissemination of antibiotic resistance. *FEMS Microbiol. Rev*. 2011;35: 872-900.

Zarrilli, R; Tripodi, MF; Di Popolo, A; Fortunato, R; Bagattini, M; Crispino, M; Florio, A; Triassi, M; Utili, R. Molecular epidemiology of high-level aminoglycoside-resistant enterococci isolated from patients in a university hospital in southern Italy. *J. Antimicrob. Chemother*. 2005; 56:827–835.

In: *Enterococcus faecalis*
Editor: Henry L. Mack

ISBN: 978-1-63321-049-3
© 2014 Nova Science Publishers, Inc.

Chapter 6

ENTEROCOCCUS FAECALIS IN ENDODONTICS

María Gabriela Pacios, DDS and María Elena López[*], *DBC, PhD*[1]
Cátedra de Química Biológica, Facultad de Odontología, Universidad Nacional de Tucumán, Argentina

ABSTRACT

Enterococcus faecalis is a persistent agent that frequently causes infection of the tooth root canal and failure of endodontic treatments. The infection is hard to be treated. Chemo mechanical cleaning and shaping of the root canal can greatly reduce the number of bacteria. However, the use of intracanal medications to disinfect the root canal system has been advocated to enhance the success of the treatment.

Calcium hydroxide is widely used as an intracanal medicament in endodontic therapy. The in vitro calcium hydroxide action and its vehicles evaluated against Enterococcus faecalis showed that this bacterium had the higher inhibition zones with calcium hydroxide + p-monochlorophenol; calcium hydroxide + p-monochlorophenol-propylene glycol pastes and 2% chlorhexidine gluconate in comparison to other intracanal medicaments. The vehicle used to prepare the calcium hydroxide paste might contribute to its antibacterial action. Chlorhexidine gluconate gel used alone, and camphorated p-monochlorophenol and camphorated p-monochlorophenol- propylene glycol as vehicles of calcium hydroxide, could be recommended, in an antimicrobial sense.

Since chlorhexidine gluconate may be used in different forms, another experience demonstrated the best of them to evidence efficacy to eliminate the most resistant intracanal bacteria. Maxillary anterior human teeth were prepared, sterilized and infected with Enterococcus faecalis for 3 days. Specimens were filled with calcium hydroxide + distilled water, 2% chlorhexidine gel and calcium hydroxide + 2% chlorhexidine (aqueous solution) and incubated at 37°C. At different times the dressings were removed and the teeth were immersed in Brain Heart Infusion broth. Enterococcus faecalis growth was evaluated by monitoring turbidity of the culture medium. Specimens were observed by scanning electron microscopy. Chlorhexidine gel was effective in eliminating Enterococcus faecalis at day 1. Calcium hydroxide + distilled water and calcium hydroxide + chlorhexidine aqueous solution showed no antimicrobial effect on Enterococcus faecalis. Scanning electron microscopy observations evidenced these results. Chlorhexidine gel was the only effective intracanal medicament against Enterococcus faecalis.

A common clinical problem in Endodontics which is the re-infection of the root canal by Enterococcus faecalis, and ultimately, the failure of the endodontic treatment, may be satisfactory solved by the use of 2% chlorhexidine gluconate gel.

INTRODUCTION

The success of the tooth root canal treatment is dependent on mechanical preparation, irrigation, microbial control, and complete obturation of the root canal. However, the occurrence of persistent root canal infection is a common clinical problem in Endodontics.

Chemomechanical preparation cannot eliminate all the bacteria from the root canal system. Byström et al. (1985) demonstrated that Instrumentation and irrigation will eliminate bacteria in only approximately 50% of root canals. The remaining bacteria may multiply during the period between appointments, in cases where the canal is not dressed with a disinfectant. However, when biomechanical preparation is combined with placement of an antimicrobial dressing before root canal obturation, bacteria can be more effectively eliminated (Sjögren et al., 1991).

Dentinal tubules may act as an important reservoir for bacteria that can lead to re-infection of the root canal and ultimately the failure of endodontic treatment (Love, 2001). Numerous studies have shown that persistent endodontic infections are often caused by Enterococcus faecalis. Molander et al. (1998) examined the microbiological status of the roots of filled teeth with periradicular lesions and found E. faecalis in 32% of the investigated teeth. Peciuliene et al. (2000), found E. faecalis in 71% of culture-positive teeth with apical periodontitis requiring retreatment. E. faecalis seems to be highly resistant to the medications used during treatment. This bacterium is easily destroyed when grown in vitro (Almyroudi et al., 2002; Gomes et al., 2001) but becomes resistant in the environment of the root canal system. There are several virulence factors which makes E. faecalis more resistant. It may undergo changes in the root canal system, possibly activating a virulence factors including adherence to host cells, expression of proteins to ensure cell survival as a result of altered environmental nutrient supply, the ability to compete with other bacterial cells, to alter the host's response and environment and alternatively, it may form a biofilm (Distel et al., 2002).

Calcium hydroxide (Ca[OH]$_2$), widely used as an intracanal medicament in Endodontic therapy. Ca(OH)$_2$ causes its antimicrobial effect through the inhibition of bacterial enzymes and the activation of tissue enzymes, such as alkaline phosphatase, causing a mineralizing effect. Its high pH inhibits bacterial metabolism, growth, and cellular division. It also alters the integrity of the cytoplasmic membrane by disruption of organic components and the transport of nutrients.

However Ca(OH)$_2$ depth of penetration in the dentinal tubules is under scrutiny and several bacterial species, including E. faecalis, are reportedly resistant to its effects. Safavi et al., (1990), and Siqueira and Uzeda (1996), suggested that Ca(OH)$_2$ preparations placed in root canals for extended periods were unable to eliminate E. faecalis.

Different vehicles have been added to Ca(OH)$_2$ in an attempt to enhance its antimicrobial activity, biocompatibility, ionic dissociation, and diffusion.

Chlorhexidine (CHX) has a broad spectrum against Gram-positive and Gram-negative bacteria. It is biocompatible and has the ability to adsorb to dental tissues and mucous membranes, with prolonged gradual release at therapeutic levels (substantivity). The antimicrobial effect of CHX is caused by the cationic molecule that binds to the negatively-charged bacterial cell walls, thereby altering the cell's osmotic equilibrium. CHX has been used in Endodontics as an irrigant and intracanal medicament.

Respect to CLX, Heling et al., (1992), demonstrated that when it was used as an intracanal medicament, it was more effective than Ca(OH)$_2$ in eliminating E. faecalis infection within dentinal tubules. Sukawat and Srisuwan (2002), demonstrated that Ca(OH)$_2$ + 0.2% CHX has the same antimicrobial effect as Ca(OH)$_2$ mixed with distilled water.

Camphorated p-monochlorophenol (CMCP) is a phenolic compound used as a disinfectant in root canal treatment. Some studies have indicated that Ca(OH)$_2$ + CMCP is the best intracanal dressing.

Propylene glycol (PG) is a clear, colorless, and odorless liquid. The advantage of this substance is its consistency, which improves the handling qualities of the paste.

There are different methods to evaluate antimicrobial action of the medicaments. Agar Diffusion Test (ADT) was used to evaluate the antibacterial action of vehicles alone and mixed with Ca(OH)$_2$ against E. faecalis. In this study a suspension of 0.1 mL of E. faecalis ATCC 29212 was inoculated into 20 mL of Mueller Hinton agar (Oxoid, UK) at 46°C, vortexed, and placed in Petri plates. Wells 4 mm deep and 4 mm in diameter were made in the agar, and completely filled with the follows medicaments: (a) Distilled water (DW); (b) 0.2% CHX solution;(c) 1% CHX solution; (d) 2% CHX solution; (e) 2% CHX gel; (f) PG; (g) PMCF; (h) PMCF-PG. The following pastes were also used: (a) Ca(OH)$_2$ + DW; (b) Ca(OH)$_2$ + 0.2% CHX; (c) Ca(OH)$_2$ + 1% CHX; (d) Ca(OH)$_2$ + 2% CHX; (e) Ca(OH)$_2$ + 2% CHX gel; (f) Ca(OH)$_2$ + PG; (g) Ca(OH)$_2$ + PMCF; (h) Ca(OH)2 + PMCF-PG .

Plates were kept for 2 h at room temperature to allow the diffusion of the medicaments through the agar, and then incubated at 37°C for 48 h. Plates with bacteria and no medicament were maintained under identical incubation conditions and considered positive controls. The diameters of the bacterial inhibition zones for each medicament, expressed in mm, were recorded. The mean values of the zones of inhibited bacterial growth of each Ca(OH)$_2$ vehicles are shown in Table 1.

Table 1. Means (+/-SE) of the diameters (in mm) of E. faecalis inhibition zones produced by the medicaments in the Agar Diffusion Test

Vehicles	E faecalis ATCC 29212
0.2% chlorhexidine	0.0+/-0.0[a]
0.2% chlorhexidine	20.3+/-1.4[b]
1% chlorhexidine	24.0+/-0.5[c]
2% chlorhexidine	26.0+/-0.5[c]
2% chlorhexidine gel	25.6+/-0.3[c]
propylene glycol	0.0+/-0.0[a]
camphorated paramonochlorophenol	20.0+/-1.1[b]
camphorated paramonochlorophenol + propylene glycol	18.3+/-0.8[b]

Different letters indicate significant difference.

E. faecalis was not inhibited by DW or PG. One percent and 2% CHX, and 2% CHX gel, showed significantly larger zones of inhibition than 0.2% CHX, CMCP, and CMCP–PG.

The means of the zones of inhibition of bacterial growth for each of the pastes are shown in Table 2. $Ca(OH)_2$ + CMCP and $Ca(OH)_2$ + CMCP–PG showed higher but not statistically-different inhibition zones than $Ca(OH)_2$ + CHX at all concentrations and both forms. No inhibition with $Ca(OH)_2$ + DW and $Ca(OH)_2$ + PG was observed with E. faecalis.

Table 2. Means (+/-SE) of the diameters (in mm) of E. faecalis inhibition zones produced by calcium hydroxide pastes in the Agar Diffusion Test

Calcium Hydroxide Pastes	E faecalis ATCC 29212
$Ca(OH)_2$ + Distilled Water	0.0+/-0.0[a]
$Ca(OH)_2$ + 0.2% chlorhexidine	18.6+/-1.4[b]
$Ca(OH)_2$ + 1% chlorhexidine	19.0+/-1.1[b]
$Ca(OH)_2$ + 2% chlorhexidine	20.0+/-0.5[b]
$Ca(OH)_2$ + 2% chlorhexidine gel	20.3+/-1.4[b]
$Ca(OH)_2$ + propylene glycol	0.0+/-0.0[a]
$Ca(OH)_2$ + camphorated paramonochlorophenol	21.6+/-0.8[b]
$Ca(OH)_2$ + camphorated paramonochlorophenol + propylene glycol	23.3+/-0.8[b]

Different letters indicate significant difference.

The zones of inhibition of a medicament depend on its toxicity against the bacteria tested and the diffusion capacity of the medicament in the medium. Substances that act by means of extreme pH values, such as bases or acids, might have their activities altered by the buffering ability of the media.

$Ca(OH)_2$ has a low solubility, does not diffuse well, and requires a long time to alkalinize the culture medium. However, in a clinical situation, the buffer ability of blood, tissue fluids, and dentin might exercise the same effects. This is why special care was taken to keep the plates at room temperature for 2 h to allow the diffusion of the medicaments through the agar, after which they were incubated under appropriate conditions. Agar diffusion requires a careful standardization of inoculum density, medium content, agar viscosity, and the size and the number of specimens per plate.

Camphorated p-monochlorophenol, CMCP–PG, and CHX at different concentrations were effective against all microorganisms. CMCP and CMCP-PG showed the same behavior. CHX efficacy was concentration dependent, and no difference was observed according to the form of presentation, as 2% CHX gel showed the same results as the 2% CLX solution. CMCP–PG and CMCP cannot be used alone as topical intracanal medicament dressings because of their cytotoxic effect, while CHX in gel form can. Enterococcus faecalis is susceptible to CHX, and a prolonged application of CHX might prove to be advantageous when used in retreatment cases.

Calcium hydroxide plays an important role in endodontics, because of its ability to induce hard tissue formation, its antibacterial action, and its tissue-dissolving capability. In addition, acting as a physical barrier, the $Ca(OH)_2$ dressing might both prevent root canal reinfection and interrupt the nutrient supply to the remaining bacteria. $Ca(OH)_2$ + DW and $Ca(OH)_2$ +

PG failed to eliminate Enterococcus faecalis, as it tolerates very high pH values of approximately 11.5.

Ca(OH)2 + CMCP and Ca(OH)2 + CMCP–PG showed large zones of inhibition against all the bacteria tested. These results are in agreement with Gomes et al. who demonstrated that the addition of glycerin to Ca(OH)2 + CMCP enhances its antimicrobial action, because glycerin helps increase the diffusibility of the paste. In this study, the addition of PG did not affect the antibacterial action. These results are also in agreement with other authors, who demonstrated that Ca(OH)$_2$ + CMCP is the most effective intracanal medication. In the present study, Ca(OH)$_2$ + CMCP and Ca(OH)$_2$ + CMCP–PG showed larger zones of inhibition than Ca(OH)2 alone, suggesting that CMCP and CMCP–PG increase its antimicrobial activity. It has been reported that CMCP has a cytotoxic effect to target the periodontal ligament cells by inhibiting cell viability and proliferation.24 This property limits its clinical use. However, periapical tissues had a good response to Ca(OH)$_2$ mixed with CMCP. Leonardo et al. suggested that the calcium p-monochlorophenolate slowly releases CMCP, and its concentration is not significant enough to be cytotoxic. In addition, the denaturing effect of Ca(OH)$_2$ on connective tissue might prevent the tissue penetration of CMCP, reducing its cytotoxicity.

Pastes prepared with Ca(OH)$_2$ + CHX were effective against all the tested bacteria. However, the inhibition zones produced by Ca(OH)$_2$ + CHX pastes at different concentrations and in solution and gel form were smaller than those zones obtained using CHX alone. These results are consistent with those demonstrated by other authors. In the Ca(OH)$_2$ + CHX mixtures, the inhibition of the antimicrobial activity of CHX, most likely caused by the high pH and the alkaline buffering capacity of the suspension, was observed. The reason for this reduced efficacy could be related to the deprotonation of the biguanide at pH >10, and thus, a markedly reduced solubility and altered interaction with the bacterial surfaces due to the change in the charge of the molecule.

Calcium hydroxide cannot be considered a universal intracanal medication, because it is not equally effective against all microorganisms in the root canal. CHX has several properties that indicate that it could be a suitable alternative to Ca(OH)2 as an intracanal medication. Applied in gel form, it might solve the irregularities of the root canal, it has a broad antimicrobial spectrum, is biocompatible with periapical tissues, and is absorbed onto hydroxyapatite of the dentin and could be subsequently be released.

Our results suggest that the vehicle used to prepare the Ca(OH)$_2$ paste might contribute to its antibacterial action, with CMCP and CMCP–PG being the best antibacterial vehicles, while 2% CHX gel alone could also be recommended as an intracanal medicament in multiple appointment root canal therapy for necrotic and retreatment cases.

For the other hand we evaluate over time, the antimicrobial effect of Ca(OH)$_2$ + distilled water, Ca(OH)$_2$ + CHX and CHX gel against E. faecalis present in the root canal of human teeth. Seventy eight human maxillary anterior teeth were used throughout. The clinical crowns were removed, the canals were instrumented with the step-back technique, after each instrument was used, the canals were irrigated with 1 ml of 1% NaOCl and the smear layer was removed by an ultrasonic bath in 17% EDTA for 5 min followed by an ultrasonic bath in 5.25% NaOCl for 5 min. Dentin specimen was contaminated with a biofilm-forming activity bacterial strain E. faecalis ATCC 29212 for 3 days. After the infection period, the specimens were dried with sterile gauze, sterile paper point and were divided into 5 test groups of 15 specimens each, according to the intracanal medications used as follows:

Group 1: Ca(OH)$_2$ + distilled water (1:1, w/v)
Group 2: Ca(OH)$_2$ + 2% CHX gluconate solution (1:1, w/v)
Group 3: 2% CHX gel
Group 4: Positive control (infected, no medication)
Group 5: Negative control (uninfected, no medication)

The Ca(OH)$_2$ pastes were placed using k-file #50 until the canal was fully packed. The paste was further condensed with the aid of absorbent paper points and vertical pluggers (Estrela et al., 2002). The 2% CHX gel was placed employing the syringe and needle supplied. The coronal and apical parts of each specimen were sealed with Cavit.

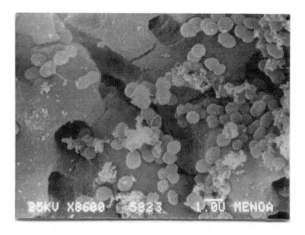

Figure 1.

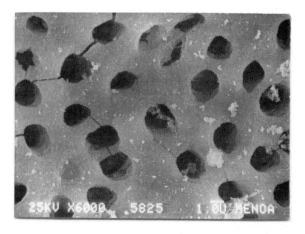

Figure 2.

Then, the external surfaces were disinfected for 10 min with 5.25% NaOCl. The specimens were placed in Petri dishes on sterile cotton wool dampened with saline to create a humid environment. The dishes were placed in an incubation jar at 37°C for 1, 2, 4, 7 and 14 d. After each time three specimens of each test group were selected. The Cavit was eliminated and the medication was removed with file #50, irrigating with 20 ml of sterile distilled water.

The specimens were then immersed in 5 ml of BHI and incubated with agitation at 50 rpm and 37°C for 48 h. E. faecalis growth was evaluated by monitoring turbidity of the culture medium and confirmed by culture on Streptococcus Faecalis (SF) medium (Becton Dickinson and Company, Detroit, MI, USA) at 37°C for 48 h and by Gram stains. Scanning Electron Microscopic (SEM) Analysis Specimens for SEM analysis were processed simultaneously with the specimens for the in vitro bacteriological assay.

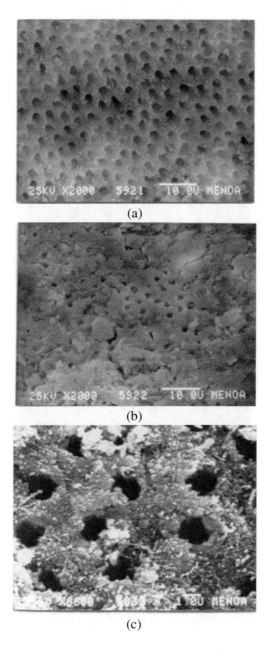

Figure 3.

Figure 4.

Our results demonstrated that 2% CHX gel was effective in eliminating E. faecalis at 24 h, while Ca(OH)$_2$ + distilled water and Ca(OH)$_2$ + 2% CHX showed no antimicrobial effect on E. faecalis from 1 to 14 d. SEM analysis showed E. faecalis in the dentinal tubules of human teeth after the infection period (Figure 1). The negative control showed no bacterial colonization in the root canal until 14 d (Figure 2).

Specimens treated with 2% CHX gel showed no bacteria at 1 d (Figure 3a), at 7 d (Figure 3b) and at 14 d (Figure 3c).

(a)

(b)

(c)

Figure 5.

In teeth treated with Ca(OH)2 + distilled water bacteria were observed onto the root canal at 1 d (Figure 4a), 7 d (Figure 4b) and 14 d (Figure 4c).

They were also observed in teeth treated with Ca(OH)2 + 2% CHX at 1 d (Figure 5a), 7 d (Figure 5b) and 14 d (Figure 5c).

E. faecalis, a facultative anaerobic Gram-positive coccus, has been implicated in persistent root canal infection (Molander et al., 1998) and blamed for causing failure of root canal treatment. It is one of the most resistant microorganisms in root canal flora. In this study an ATCC strain previously used in vitro studies to evaluate the antibacterial effects of intracanal dressing (Distel et al., 2002; Estrela et al., 1999; Sukawat and Srisuwan, 2002) was selected.

Ca(OH)$_2$, widely used for apexification and pulp capping procedure, is nowadays widely used as an intracanal medicament in endodontic therapy. Its high pH inhibits essential enzymes activities for metabolisms, growth, and cellular division of bacteria. The influence of pH alters the integrity of the cytoplasm membrane by disrupting organic components and transport of nutrients (Safavi et al., 1990). Ca(OH)$_2$ was chosen for this study because it is currently the most popular intracanal medication. However, there is some controversy about its efficacy against E. faecalis. CHX is a cationic bisguanide that seems to act by adsorption to the cell wall of the microorganisms, causing leakage of intracellular components. At low CHX concentrations, small molecular weight substances will leak out, specifically potassium and phosphorous, resulting in a bacteriostatic effect. At higher concentrations, CHX has a bactericidal effect due to precipitation and/or coagulation of the cytoplasm, probably caused by protein cross-linking (Gomes et al., 2001; White et al., 1997). The tested CHX was used in the gel form (natrozole) to facilitate its application and prolong its retention in the root canal. Natrozole is a biocompatible carbon polymer, a water-soluble substance that can be completely removed from the root canal with a final flush of distilled water. It also has no antibacterial effect.

E. faecalis were observed in the root dentin of a positive control since the first day of infection until the end of the experiment, and was confirmed alive through its growth in a specific media for E. faecalis. Waltimo et al., (2000), demonstrated that a very short incubation period was sufficient for the growth of E. faecalis in the dentinal tubules. Behnen et al., (2001), demonstrated its ability to penetrate into the dentinal tubules of bovine teeth after 24 h incubation. Evans et al., (2002), infected bovine root canals in 5 d. However, our study was not performed with the aim to demonstrate either adhesion or penetration of bacteria into tubules. It was just necessary to find at list one bacteria alive that can multiply in a specific growth media, in order to evidence the effectiveness of the antibacterial effect of an intracanal medicament. That is, the merely presence of bacteria makes unnecessary any quantitative analysis.

In this study 2% CHX gel was effective in the elimination of E. faecalis within dentinal tubules, while Ca(OH)$_2$ + distilled water and Ca(OH)2 + CHX were ineffective in eliminating the bacteria. Gomes et al., (2003), also demonstrated that 2% CHX gel alone was more effective against E. faecalis than Ca(OH)2. In our study the Ca(OH)2 and CHX combination showed the same antimicrobial effect as mixed with distilled water. These results are in agreement with those of Sukuwat and Srisuwan (2002), and Haenni et al., (2003), CHX did not have a synergic effect on the antibacterial action of Ca(OH)2. Quite the opposite, the Ca(OH)2-CHX mixture, showed inhibition of the antibacterial activity of CHX. The possible reason for this reduced efficacy may be the deprotonation of the biguanide at pH > 10 and hence, a markedly reduced solubility and altered interaction with bacterial surfaces due to the change in the charge of the molecule. Controversially many authors demonstrated that Ca(OH)2 with CHX was more effective in eliminating E. faecalis from dentinal tubules than Ca(OH)2 with distilled water (Basrani et al., 2003; Delgado et al., 2010; Evans et al., 2003; Lin et al., 2003).

In human teeth, Siqueira and Uzeda (1993), and Ørstavik and Haapasalo (1990), reported that Ca(OH)2 was unable to completely eliminate E. faecalis from dentinal tubules after 10 d. Conversely, Han et al., (2001), demonstrated that aqueous Ca(OH)2 paste was effective in the elimination of E. faecalis in the dentinal tubules after 7 d. Almiroundi et al., (2002), demonstrated that Ca(OH)2 combined with distilled water killed E. faecalis within the dentinal tubules at 3 d and 8 d, but failed to do so at 14 d; the CHX gel and Ca(OH)2 combined with CHX was effective in eliminating E. faecalis inside dentinal tubules, as reported by the authors. These studies involved incubating dentin shavings or dentin powder after the action of the intracanal medicament. In our study, teeth were directly incubated following the methodology of Estrela et al., (1999), who also demonstrated that Ca(OH)2 failed to eliminate E. faecalis from dentinal tubules after 7 d.

The antimicrobial action of Ca(OH)$_2$ is related to the release of OH- ions in an aqueous environment and therefore depends on the availability of OH- ions in the solution and the ability of these ions to diffuse through dentin and pulpar tissue remnants to reach sequestered bacteria. Dentin itself has buffer properties for the base, so it might decrease the pH effect of the Ca(OH)$_2$. Evans et al., (2002), verified that the proton pump of the E. faecalis cell, which carries protons to the interior of the cell acidifying the cytoplasm, is important for the survival of the microorganism in an alkaline environment. Presumably, when the alkalinity of the environment reaches pH 11.5 or higher, this "lifeguard" mechanism is brought into action. This mechanism, associated to the dentin buffer properties, can explain the resistance of E. faecalis to Ca(OH)$_2$ alone.

CONCLUSION

The use of an antimicrobial agent as an intracanal medication between appointments may increase the chances of a successful endodontic treatment by reducing residual bacteria in the root canal system. Results expressed in this chapter indicate that 2% CHX gel may exert an important role in the eradication of endodontic infection by E. faecalis associated with teeth that were refractory to the conventional endodontic therapy.

REFERENCES

Almyroudi, A, Mackenzie, D, McHugh, S, Saunders, WP. (2002) The effectiveness of various disinfectants used as endodontic intracanal medications: an *in vitro* study. *Journal of Endodontics*. 28, 163-167.

Basrani, B, Tjäderhane, L, Santos, M, Pascon, E, Grad, H, Lawrence, HP, Friedman, S. (2003) Efficacy of chlorhexidine – and calcium hydroxide – containing medicaments against *Enterococcus faecalis in vitro*. *Oral Surgery, Oral Medicine, Oral Pathology, Oral Radiology and Endodontics*. 96, 618-624.

Behnen, MJ, West, LA, Liewehr, FR, Buxton, TB, McPerson, JC. (2001) Antimicrobial activity of several calcium hydroxide preparations in root dentin. *Journal of Endodontics*. 27, 765-767.

Byström, A, Claesson, R, Sundqvist, G. (1985) The antibacterial effect of camphorated paramonochlorophenol, camphorated phenol, and calcium hydroxide in the treatment of infected root canals. *Endodontics and Dental Traumatology*. 1, 170-175.

Delgado, RJ, Gasparoto, TH, Sipert, CR, Pinheiro, CR, Moraes, IG, García, RB, Bramante, CM, Campanelli, AP, Bernardinelli, N. (2010) Antimicrobial effects of calcium hydroxide and chlorhexidine on *Enterococcus faecalis*. *Journal of Endodontics.* 36, 1389-1393.

Distel, JW, Hatton, JF, Gillespie, MJ. (2002) Biofilm formation in medicated root canals. *Journal of Endodontics.* 28, 689-693.

Estrela, C, Pimenta, FC, Ito, YI, Bamman, LL. (1999) Antimicrobial evaluation of calcium hydroxide in infected dentinal tubules. *Journal of Endodontics.* 25, 416-418.

Estrela, C, Mamede Neto, I, Lopes, HP, Estrela, CR, Pécora, JD. (2002) Root canal filling with calcium hydroxide using different techniques. *Brazilian Dental Journal.* 13, 53-56.

Evans, MD, Davies, JK, Sundqvist, G, Figdor, D. (2002) Mechanisms involved in the resistance of *Enterococcus faecalis* to calcium hydroxide. *International Endodontic Journal.* 35, 221-228.

Evans, MD, Baumgartner, JC, Khemaleelakul, SU, Xia, T. (2003) Efficacy of calcium hydroxide: chlorhexidine paste as an intracanal medication in bovine dentin. *Journal of Endodontics.* 29, 338-339.

Gomes, BPFA, Feraz, CCR, Vianna, ME, Berber, VB, Teixeira, FB, Souza-Filho, FJ. (2001) In vitro antimicrobial activity of several concentrations of sodium hypochlorite and chlorhexidine gluconate in the elimination of *Enterococcus faecalis*. *International Endodontic Journal.* 34, 424-428.

Gomes, BP, Souza, SF, Ferraz, CC, Teixeira, FB, Zaia, AA, Valdrighi, L, Souza-Filho, FJ. (2003) Effectiveness of 2% chlorhexidine gel and calcium hydroxide against *Enterococcus faecalis* in bovine root dentine *in vitro*. *International Endodontic Journal.* 36, 267-275.

Haenni, S, Schmidlin, PR, Mueller, B, Sener, B, Zehnder, M. (2003) Chemical and antimicrobial properties of calcium hydroxide mixed with irrigating solutions. *International Endodontic Journal.* 36, 100-105.

Han, GY, Park, SH, Yoon, TC. (2001) Antimicrobial activity of $Ca(OH)_2$ containing pastes with *Enterococcus faecalis in vitro*. *Journal of Endodontics.* 27, 328-332.

Heling, I, Steinberg, D, Kening, S, Gavrilovich, I, Sela, MN, Friedman, M. (1992) Efficacy of a sustained-release device containing chlorhexidine and $Ca(OH)_2$ in preventing secondary infection of dentinal tubules. *International Endodontic Journal.* 25, 20-24.

Leonardo, MR. (1994) Preparación biomecánica de los conductos radiculares. In: Leonardo, MR, Leal, JM, editors. Endodoncia: Tratamiento de los conductos radiculares. *Argentina: Buenos Aires.* 296-320.

Lin, Y, Mickel, AK, Chogle, S. (2003) Effectiveness of selected materials against *Enterococcus faecalis*: Part 3. The antimicrobial effect of calcium hydroxide and chlorhexidine on *Enterococcus faecalis*. *Journal of Endodontics.* 29, 565-566.

Love, RM. (2001) *Enterococcus faecalis* – a mechanism for its role in endodontic failure. *International Endodontic Journal.* 34, 399-405.

Molander, A, Reit, C, Dalhén, G, Kvist, T. (1998) Microbiological status of root-filled teeth with periodontitis. *International Endodontic Journal.* 31, 1-7.

Ørstavik, D, Haapasalo, M. (1990) Disinfection by endodontic irrigants and dressings of experimentally infected dentinal tubules. *Endodontics and Dental Traumatology.* 27, 218-222.

Peciuliene, V, Balciuniene, I, Eriksen, H, Haapasalo, M. (2000) Isolation of *Enterococcus faecalis* in previously root-filled canals in Lithuanian population. *Journal of Endodontics.* 26, 593-595.

Safavi, KE, Spångberg, LS, Langeland, K. (1990) Root canal dentinal tubule disinfection. *Journal of Endodontics.* 16, 207-210.

Siquiera, JF Jr, Uzeda, M. (1996) Disinfection by calcium hydroxide pastes of dentinal tubules infected with two obligate and one facultative anaerobic bacteria. *Journal of Endodontics.* 22, 674-676.

Sjögren, U, Fidgor, D, Spångberg, L, Sundqvist, G. (1991) The antimicrobial effect of calcium hydroxide as a short-term intracanal dressing. *International Endodontic Journal.* 24, 119-125.

Stuart, CH, Schwartz, SA, Beeson, TJ, Owatz, CB. (2006) *Enterococcus faecalis*: Its role in root canal treatment failure and current concepts in retirement. *Journal of Endodontics.* 32, 93-98.

Sukawat, C, Srisuwan, T. (2002) A comparison of the antimicrobial efficacy of three calcium hydroxide formulations on human dentin infected with *Enterococcus faecalis*. *Journal of Endodontics.* 28, 102-104.

Waltimo, TMT, Ørtavik, D, Siren, EK, Haapasalo, MPP. (2000) *In vitro* yeast infection of human dentin. Journal of Endodontics.26, 207-209.

White, RR, Hays, GL, Janer, LR. (1997) Residual antimicrobial activity after canal irrigation with chlorhexidine. Journal of Endodontics. 23, 229-231.

Zerella, JA, Fouad, AF, Spangberg, LS. (2005) Effectiveness of calcium hydroxide and chlorhexidine digluconate mixture as disinfectant during retreatment of failed endodontic cases. *Oral Surgery, Oral Medicine, Oral Pathology, Oral Radiology and Endodontic*s. 100, 756-761.

In: *Enterococcus faecalis*
Editor: Henry L. Mack

ISBN: 978-1-63321-049-3
© 2014 Nova Science Publishers, Inc.

Chapter 7

MOLECULAR CHARACTERIZATION OF NATURAL DAIRY ISOLATES OF *ENTEROCOCCUS FAECALIS* AND EVALUATION OF THEIR ANTIMICROBIAL POTENTIAL

Katarina Veljović[*]*, Amarela Terzić-Vidojević, Maja Tolinački, Sanja Mihajlović, Goran Vukotić, Natasa Golić and Milan Kojić*
Laboratory for Molecular Microbiology, Institute of Molecular Genetics and Genetic Engineering, University of Belgrade, Serbia

ABSTRACT

Due to their ability to survive adverse conditions, **enterococci are widespread** in nature and can be found in milk, dairy products and human and animal gastrointestinal tracts. Still, the use of enterococci in food preparation is controversial, since they have traditionally been branded as indicators of faecal contamination and their role in food spoilage is well known. However, some enterococcal strains exhibit antimicrobial effects and have probiotic potential, contributing to the improvement of the general state of health. For that reason, we have analyzed natural isolates of *Enterococcus faecalis* originating from various dairy products manufactured in rural households located in the mountains of Serbia. Genotyping analysis of selected enterococci showed high diversity among the isolates. The antimicrobial activity of the isolates showed a great effect on the number of pathogenic and non-pathogenic strains, including *L. monocytogenes, L. innocua, E. coli, Pseudomonas sp.,* and *Candida pseudotropicalis*. Furthermore, analysis of the presence of known bacteriocin encoding genes showed that the genes for various enterocins were present. Although in some strains more than one enterocin gene was detected, there was no correlation between the number of enterocin genes and the antimicrobial spectrum. Nevertheless, in order to characterize the strains that could be safely used as starter cultures in functional food, the frequency of virulence determinants and antibiotic resistance, as well as the synthesis of biogenic amines, was analyzed. The results show that the presence of virulence determinants and antibiotic resistance is strain

---

[*] Corresponding author: PhD Katarina Veljovic. E-mail: katarinak76@yahoo.com, katarinav@ imgge.bg.ac.rs.

dependent and region specific. In addition, a large percentage of the strains have the ability to decarboxylate tyrosine and other amino acids. Such capacity for decarboxylation of amino acids limits the use of the strains in the food industry. Based on these results, it can be concluded that enterococci isolated from animal food must be viewed with particular caution because they are reservoirs of genes for antibiotic resistance and virulence.

INTRODUCTION

Thanks to their ability to survive adverse conditions, bacteria belonging to the genus *Enterococcus* are widely distributed in nature (Franz et al., 2003). Enterococci comprise a significant part of the indigenous microflora of the gastrointestinal (GI) tracts of humans and animals and are found to be very important to maintaining the balance of gut microorganisms (Leroy et al., 2003). Once released into the environment through human and animal faeces, enterococci are able to inhabit different ecological niches such as soil, surface water and various plants (Giraffa, 2003).

Due to contamination of the surrounding environment, enterococci enter the milk of domesticated animals through grazing.

Since they have the ability to survive various disadvantageous conditions, such as extremes in pH, temperature and salinity, enterococci withstand a variety of technological conditions during food preparation, and can be found in fermented dairy and meat products, as well as other types of alimentary pro-ducts (Giraffa, 2003).

As part of non-starter cultures, together with other lactic acid bacteria (LAB), enterococci are generally involved in food fermentation (Franz et al., 1999). Due to their proteolytic and lipolytic activities, degradation of citrate, and production of diacetyl and acetoin, enterococci directly contribute to the formation of the specific flavour of the final fermented product, and have an important role in cheese ripening.

In addition, enterococci synthesize proteins and other bioactive substances which react with the food components and support other significant biochemi-cal changes associated with biopreservation (Giraffa, 2003).

In addition, enterococci have been used for years as probiotics, contributing to the improvement of the general state of health of an organism (Tannock and Cook, 2002).

DAIRY NATURAL ISOLATES OF *EN. FAECALIS*

A large number of strains isolated from milk and dairy products from the Western Balkan Countries (WBC) region belong to the species enterococci (Terzic et al., 2007, 2014; Veljovic et al., 2007, 2009; Jokovic, 2008; Golic et al., 2013). It is known that some parts of the WBC represent specific ecological sites, such as the mountains Zlatar, Radan and Stara Planina, Vlasina and other locations where traditional food produced in households represents a potential source of bacterial strains of technological interest (Topisirovic et al., 2006). Such specific diversity has an impact on the formation of the specific microflora found in milk. The natural food isolates of enterococci originate from fresh and fermented products manufactured in households without the addition of starter cultures. Several studies have suggested that dairy food strains of enterococci have a positive influence on the ripening of traditional cheeses (Giraffa, 2003; Foulquie Moreno et al., 2006). These data correspond to the fact that enterococci are considered essential for cheese flavour in most Southern European countries. In contrast, most of the studies from Northern European countries are more focused on the negative aspects of enterococci (Ogier and Serror, 2008). These divergent viewpoints may simply be due to cultural differences.

Among enterococci, the most common species isolated from raw milk and fermented milk products from the WBC region are *En. faecalis* and *En. faecium*. Although the literature data demonstrate the dominance of *En. faecalis*, among other enterococci, in dairy products (Giraffa et al., 2003; Nieto-Arribas et al., 2011; Jamet et al., 2012), an analysis of *Enterococcus* species in fermented products of the WBC region in general revealed that 97 out of 501 enterococci strains are *En. faecalis* (19.36%) (unpublished data). However, in some specific products *En. faecalis* were found to be dominant. For example, a total of 68.75% of cocci isolated from white pickled Zlatar cheese (originating from Zlatar Mountain, Serbia) (Veljovic et al., 2007) and 24% in BGAL2 cheese (Aleksinac, Serbia) belongs to *En. faecalis* species, as well as 55% in fresh soft ZGZA8 cheese (originating from Zagorje, Croatia), and 41% in ZGPR2 cheese (Prigorje region, Croatia) (Golic et al., 2013).

Considering that the genus *Enterococcus* includes 52 species (http://www.bacterio.cict.fr/bacdico/ee/Enterococcus.htlm, March, 1st 2014), *En. faecalis* is now known to have a special significance and role in food fermentation, although it has also been associated with human diseases (Franz et al., 2003).

MOLECULAR CHARACTERISATION OF *EN. FAECALIS* NATURAL DAIRY ISOLATES

There is a great interest in *En. faecalis* strains isolated from milk and dairy products, especially in the Mediterranean countries. For decades, enterococci were used in the dairy industry for production of different types of cheeses, especially as part of starters or "widgets", contributing to the improvement of sensory characteristics of the cheeses during ripening. Proteolytic and lipolytic activity in milk, followed by the synthesis of volatile compounds, such as lactate, citrate, diacetyl, acetoin, ethanol, and of course, the synthesis of bacteriocins, are some of the technological characteristics which are most important for starter cultures (Asteri et al. 2009; Nieto-Arribas et al., 2011).

Enterococci are known as relatively weak proteolites (Suzzi et al., 2000), and among them the greatest ability for casein degradation is exhibited by *En. faecalis* strains (Sarantinopoulos et al., 2001). There are different opinions about the proteolytic activity of enterococci, but the results by Andrighetto et al. (2001) and Sarantinopoulos et al. (2001) point out the high proteolytic activity of certain *En. faecalis* strains, similar to the results obtained in our previous work (Veljovic et al., 2009). Our results showed that in the WBC region 17 out of 97 natural dairy isolates of *En. facealis* showed high proteo-lityc activity (17.53%) (unpublished data).

Interestingly, the strains *En. faecalis* BGPT1-10P and BGPT1–78 are highly active in milk (Veljovic et al., 2009).

It is supposed that the formation of curd during the experiment was caused by the remarkable proteolytic activity of both strains, considering that at the time of coagulation, in both cases, the measured pH value of the milk was almost neutral, hence curdling due to acidification activity could be excluded. According to the results of Morea et al. (1999) the degree of acidification of milk inoculated with enterococci strains depends on the origin of the strains used. The authors showed that 24 h after the inoculation of milk with enterococci isolated from Mozzarella cheese, the pH value was about 5.5. On the other hand, Suzzi et al. (2000) showed a high degree of milk acidification by *En. faecalis* isolated from an Italian "Semicotto Caprino" cheese, where the pH value of the milk after 24 h of incubation was 4.5.

En. faecalis strains have the ability to synthesize volatile compounds. Synthesis of diacetyl, acetoin from glucose, and citrate in the process of meta-bolic degradation are very important for the flavour formation of fermented dairy products. Andrighetto et al. (2001) reported the highest quantity of synthesized aromatic substances in *En. faecalis* strains.

Similarly, Nieto-Arribas et al. (2011) showed that En. faecalis strains are the best producers of different volatile compounds. Our analysis of *En. Faeca-lis* dairy isolates from the WBC region showed that 86 out of 97 strains produced acetoin (88.66%), and 49 strains produced citrate (50.52%), while 24 strains synthesized diacetyl (24.74%) (unpublished data).

Randazzo et al. (2008) point out that all isolates which have the ability to degrade organic substances could be candidates for starter cultures. The proteolytic activity of enterococci, or the ability of casein degradation, is an important feature that allows them to grow in milk, and it also has an impor-tant role in the formation of the texture and taste of cheese.

Two *En. faecalis* strains, BGPT1-10P and BGPT1-78 were distinguished from the others by performing a complete degradation of all three fractions of casein. Detailed molecular characterisation of the proteinase and determination of the molecular mass of the proteinases (approximately 29 kDa) led to the conclusion that strains BGPT1-10P and BGPT1-78 have the same type of proteinase as BGPM3 (Fira et al., 2000; Veljovic et al. 2009). At the same time it was shown that all of the strains possess gelatinase. Gelatinase activity is a trait that is significantly represented among enterococci. Strains carrying out the degradation of gelatin and casein have often been isolated from milk, cheese and other fermented products (Lopez et al., 2006).

Although the presence of gelatinase may be essential to the bacteria, the presence of gelatinases is also associated with infection, and is considered a virulence factor (Eaton and Gasson, 2001).

THE VIRULENCE POTENTIAL OF DAIRY ISOLATES OF *EN. FAECALIS*

Analyses of the basic characteristics and properties of enterococci demonstrate that enterococci exhibit a so-called dualistic effect. As opposed to all of their positive characteristics, enterococci can be regarded as indicators of undesirable contamination and may carry certain pathogenic potential. Unlike other LAB, enterococci are not microorganisms classified as Generally Recognized As Safe (GRAS), and they are traditionally branded as indicators of faecal contamination of an area (Ogier and Serror, 2008). In addition, enterococci are often involved in food spoilage (Franz et al., 1999), can cause food poisoning (Gardini et al., 2001) and contribute to the spread of antibiotic resistance via the food chain (Giraffa, 2002).

For many years it was thought that enterococci were commensals with low pathogenic potential for humans. More recently this understanding has changed, due to the growing number of studies showing the involvement of enterococci in nosocomial infections, especially in patients undergoing long-term antibiotic treatment (Franz et al. 2003; Giraffa, 2003; Kayser 2003).

Although enterococci may be associated with other pathogens and it is difficult to determine their contribution to the infection, it is known that enterococci could be drivers of urinary tract infections, bacteremia, endocar-ditis and postoperative complications after cataract surgery (Malani et al., 2002).

Virulence factors that contribute to the pathogenic potential of enterococci are virulence determinants, production of biogenic amines and antibiotic resistance. In order to exclude the enterococci with pathogenic potential from isolates for possible use in fermented food production and to prevent further transfer of the associated genes to other enterococcal strains, or other bacteria in the environment, genetic characterization of virulence factors is required (Mannu et al., 2003; Franz et al., 2003).

Virulence factors are genetically encoded features in some bacterial strains, and exhibit a pathogenic effect on mammalian tissues and/or provide resistance to specific and non-specific defence mechanisms.

As a result, virulence factors allow bacteria to act as "opportunistic patho-gens" (Giménez-Pereira, 2005).

Virulence factors are generally well known among clinical isolates of enterococci. However, studies that have been done in the last decade indicate the presence of virulence factors in food isolates as well.

Among enterococcal isolates, Franz et al. (2001) and Eaton and Gasson (2001) showed that strains of En. faecalis, generally carry more virulence factors than strains of *En. faecium*.

Lopes et al. (2006) analyzed the correlation between gelatinase activity and the presence of the *genE* gene in enterococci isolated from dairy products. However, *genE* itself is not sufficient for the expression of gelatinase activity.

The presence of other genes, including the entire *fsr* operon, is required. The results of Valenzuela et al. (2009) showed that four out of eight analysed dairy *En. faecalis* strains carry the *genE* gene, although only the strains BGPT1-10P and BGPT1-78 among them have gelatinase activity (Valenzuela et al., 2009; Veljovic et al., 2009).

Besides the analysis of technological characteristics, an analysis of the frequency of virulence determinants and antibiotic resistance, as well as the syn-thesis of biogenic amines, was performed. It is known that enterococci may carry potential virulence factors which exhibit the properties of patho-genic strains. Fortunately, these important features are strain-specific, not species-specific. For this reason, the selection of enterococci strains for use in the food industry and in medicine is based on the absence of any pathogenic properties as well as the antibiotic resistance gene (Gimenez-Pereira, 2005).

Judging by published data, little is known about the mechanisms of virulence in enterococci that contribute to pathogenesis; however, several factors have been identified as potential virulence factors (Ogier and Serror, 2008). By analyzing hemolysis it has been shown that *En. faecalis* strain BG221 shows alpha-hemolytic activity (Valenzuela et al., 2009). Beta-hemolytic enterococci are not found, therefore the analyzed enterococci have no ability for lysis of red blood cells, which is more characteristic of clinical isolates. A set of three genes of the cytolysin operon (*cylM*, *cylB* and *cylA*, responsible for maturation, transport and activation of cytolysin precursors) is only detected in strain BGPT1–10P among all analysed strains. However, two additional genes, *cylL1* and *cylL2*, are required for the expression of hemolytic activity but are not present in this strain. In contrast, the genes for hemolytic activity (the gene for enterococcal surface protein *esp*, and endocarditis enterococcal antigen, *efaA*) are found in high frequency (77.77 % and 66.66 %, respectively) in all tested *En. faecalis* strains. Both genes are involved in the adhesion of enterococci to the host cell matrix (Lowe et al., 1995).

Although it was shown that enterococci isolated from cheese may carry multiple virulence determinants (Canzek Majhenic et al., 2005) and that detection of virulence genes may indicate the virulent potential of food isolates, infections with enterococci have not been recorded.

Nevertheless, isolates from food may contribute to the spread of virulence genes by horizontal transfer.

Genes for sex pheromone, which are responsible for the transfer of genetic material, are also present in the tested enterococci. Strains with the *ccf*, *cob* and *cpd* determinants have the potential to receive the corresponding plasmids, which carry the genes for sex pheromone, and thus another virulence deter-minant. All *En. faecalis* strains carry at least one of the genes for sex phero-mones, and two *En. faecalis* strains, BGPT1-10P and BGPT1-78, have all three genes for sex-pheromone, as well as the *agg* gene, which is responsible for binding to eukaryotic cells (Valenzuela et al., 2009).

The presence of agg is always associated with the presence of genes for sex pheromones, but the presence of other determinants is never conditioned (Eatton and Gasson, 2001).

Furthermore, one of the negative aspects of enterococci in cheeses is their ability to produce biogenic amines. The results of Valenzuela et al. (2009) showed that a large percentage of *En. faecalis* strains have the ability of tyrosi-ne decarboxylation (55.55%), while strain BGPT1-10P performs decarboxyla-tion of tyrosine and ornithine. The other four strains have no ability of amino acid decarboxylation. Such capacity for amino acid decarboxylation limits the use of the strains in the food industry. Biogenic amines are organic bases with aliphatic, aromatic, and heterocyclic structures which can be found in various foods (Giraffa, 2002). They are produced by the process of decarboxylation of the corresponding amino acid, through substrate-specific enzymes, synthesized by a microorganism. Cheese may be a substrate for the production and accumulation of biogenic amines, because it has a high concentration of proteins (Giraffa, 2002). It has been shown that some *Enterococcus* species are capable of synthesizing biogenic amines (Gardini et al., 2001).

Finally, antibiotic resistance has for many years been studied in detail in *En. faecalis* and *En. faecium* species (Franz et al. 2003; Clare et al., 2003), in the clinical environment and in food isolates.

The clinical treatment of infections caused by enterococci represents serious problems for immunocompromised patients, since enterococci have become resistant to an increasing number of antibiotics (Mundy et al., 2000).

Research in the last decade indicates that enterococci from food can be a reservoir of antibiotic resistance genes (Ogier and Serror, 2008). Unlike with decarboxylase activity, all *En. faecalis* species showed a high proportion of resistance to rifampicin, ciprofloxacin, levofloxacin and quinupristin/dalfo-pristin, and a slightly lower proportion of resistance to erythromycin and tetra-cycline.

Vancomycin resistance is more related to the use of antibiotics in a clinical setting (Ribeiro et al., 2007). Only BGPT1-78 strain is resistant to vancomycin. Recent studies showed that the VanA phenotype usually prevails among clinical and dairy isolates. The dominance of this phenotype can be explained by successful gene transfer using transformable elements, such as Tn1546, without the spread of resistant strains (Ribeiro et al., 2007). The plasmids and transposons carrying the gene for erythromycin and tetracycline resistance are also common among enterococci (Murray, 1999; Peters et al., 2003).

All of the fermented products from which the tested enterococci were isolated were made in a rural setting, away from urban areas. The presence of antibiotic resistance among isolates from food varies and it is most often in a single strain, or related to a specific region. In the analysis of enterococci in European cheeses, vancomycin-resistant strains are found in the frequency range of 4% to 80% (Teuber et al., 1999). The lowest level of antibiotic resistance was shown in enterococci from Tolminc cheese (Canzek Majhenic et al., 2005). It is shown that all 97 analysed *En. faecalis* strains from WBC region were resistant to some of the tested antibiotics (unpublished data).

For instance, the work of Veljovic et al. (2013) showed that the strain BGPT1-10P was resistant to erythromycin and tetracycline. Since they are antibiotics that have found wide use in human and veterinary medicine, the molecular basis of the resistance is determined. The strain carries an *ermB* gene, which encodes B methylase, responsible for resistance to erythromycin, and a *tetM* gene RPP type, which serves to protect the ribosome, and is responsible for tetracycline resistance (Veljovic et al., 2013). Teuber et al. (1999) have shown that the resistance to tetracycline and erythromycin in 45% and 32%, respectively, occurring in *En. faecalis* species isolated from dairy products, are in correlation with the mobile genetic elements which are rich in enterococci. Huys et al. (2004) explained the multidrug resistance, discussing the relationship of the *tet* genes with transposons carrying resistance to multiple antibiotics.

ANTIMICROBIAL ACTIVITY OF FOOD ENTEROCOCCI

Although the role of enterococci in food spoilage is well known, as well as their relation to clinical infections, the fact that enterococci synthesize bacteriocins with excellent antibacterial effect makes them interesting candida-tes for food biopreservation (Giménez-Pereira, 2005).

Hence, the bacteriocins from enterococci, the so-called enterocins, have been the subject of detailed study for many years (Gálvez et al., 2007) with respect to their diversity, environmental and practical importance, their use in food biopreservation and their possible therapeutic role (Franz et al., 2007). Interest in the study of enterocins has grown since the discovery of their antimicrobial effect on Gram-positive pathogenic bacteria, vegetative cells and spores (Cleveland et al., 2001; Deegan et al., 2006). Enterocins are important due to their versatility and exceptional distribution among isolates. The diversity of enterocins is reflected in the robust nature and wide availability of enterococci, as well as their extraordinary abilities to spread and to receive genetic material (Franz et al., 2007). Efficient transfer of genetic material can be explained by the existence of a large number of diverse enterocins and the variety of enterocins produced by the same strain, as well as the phenomenon that identical enterocins can be synthesized by a variety of *Enterococcus* species (Franz et al., 2007). Among *En. faecalis* and *En. faecium* species, enterocins A, B and P are widespread.

Synthesis of enterocins can enable a particular strain to colonize certain foods, particularly fermented dairy and meat products, and thus to become the most important component of the microbial community. Enterococci may become the dominant microflora in a fermented product due to its tolerance for high temperatures, dryness and increased salinity (Franz et al., 1999). Production of enterocins in combination with a wide range of tolerance to adverse chemical and physical conditions may explain why these bacteria are so robust and why they inhabit such diverse ecological niches.

The practical importance of enterocins is reflected in the inhibitory activity of enterococci on other LAB, but also on a wide variety of pathogenic bacteria, including species of the genera *Listeria, Staphylococcus, Clostridium, Bacillus* and others. Enterocins are generally active against the species *L. monocytogenes* and *C. tyrobutyricum*, food-spoiling microorganisms associated with the "late blowing" defect of cheese (Tsakalidou et al. 1993; Centeno et al., 1996). Moreover, the use of enterocin-producing enterococci as starter cultures may contribute to the prevention of food spoilage (Ananou et al. 2010; Viedma et al. 2009).

Following the findings that enterococci prevent the growth and development of many Gram-positive bacteria, vegetative cells and spores, intensive study of enterocins has been performed. The inhibitory action on pathogenic bacteria by enterocins, or enterocin-producing strains, can prevent the decay of food, giving enterococci an important role in food biopreservation (Franz et al., 2007).

For example, enterocin AS-48 has found wide application in biopreserva-tion in various food types and is especially noted for its antilisterial and anti-staphilococcal effects (Ananou et al., 2010; Viedma et al., 2009).

The antimicrobial activity in natural dairy isolates from the WBC region showed a great effect on the number of pathogenic and non-pathogenic strains. Nineteen out of 97 analysed *En. faecalis* strains showed antimicrobial activity (19.59%) (unpublished data). Antimicrobial activity detected after treatment with protease suggests the proteinaceous nature of the bacteriocin activity. Among them, only one strain, BG221, which does not synthesize enterocin, showed antimicrobial activity against five indicator strains (Veljovic et al., 2009). It was found that this strain produces hydrogen peroxide, although its antimicrobial activity was retained after treatment with catalase. Hence, it is assumed that this strain synthesizes another organic compound with inhibitory activity.

The work of Veljovic et al. (2009) revealed that all analysed *En. faecalis* dairy strains exhibit antimicrobial activity, and showed inhibitory activity against tested indicator strains. None of the *En. faecalis* strains showed inhibitory activity against lactobacilli indicator strains, while most of the tested strains were active against enterococcal and lactococcal indicator strains (Veljovic et al., 2009). It has been shown previously that a very large number of enterocins, particularly enterocins from the class II.1., including enterocins A and P, exhibit an antilisterial effect (Giraffa, 1995). Interestingly, it was shown that a total of 66.6% of tested *En. faecalis* strains exhibit an antilisterial effect both on non-pathogenic *L. innocua* and *L. monocytogenes* (Veljovic et al., 2009), a food-related pathogenic bacteria causing listeriosis (Du Toit et al., 2000). Moreover, the presence of the genes for enterocin A or enterocin P were detected in the analysed strains (Veljovic et al., 2009; Valenzuela et al.,

2009). Therefore, all tested *En. faecalis* strains with pronounced enterocin activity, in particular to food pathogens, could be successfully used in the formulation of starter cultures as biopreservatives.

In particular, strains BGPT1-10P, BGPT1-78 and BGAZES1-5 showed antimicrobial activity against Gram-negative strain *Pseudomonas* sp. PA17, while strain BGAZES1-5 inhibits the growth of *E. coli* ATCC25922, an important food-pathogen, and therefore may be good candidates for use in starter cultures for fermented food production.

Interestingly, the results of a correlation analysis between the presence of enterocin genes and antimicrobial activity of *En. faecalis* natural isolates showed that some of the strains without antimicrobial activity possess the genes for enterocins, and in fact some of them have more than one gene for enterocin production (De Vuyst et al., 2003). Although it is possible that the strains possess very specific antimicrobial activity that was not detected in the experimental conditions used in this study, the presence of inactive enterocin genes is usually explained by the very efficient transfer of genetic material common in enterococci (Franz et al., 2007).

Moreover, there is no correlation between the number of enterocin genes present in one strain and its antimicrobial spectrum. The strain BGAR131 has four enterocin genes and exhibits antimicrobial activity against seven indicator strains, while in strain BGAZES1-5 only two enterocin genes provide broad antimicrobial activity against seven different indicator strains (Veljovic et al., 2009; Valenzuela et al., 2009). PCR with primers for different enterocins allows quick and efficient detection of the presence of known enterocin genes among the tested enterococci strains. Thus, the results revealed that the strains BGPT1-10P and BGPT1-78, which are active against 11 different indicator strains, carry the *entL* gene (Veljovic et al., 2013).

CONCLUSION

Given the status of enterococci, an approach that emphasizes the need to study each strain individually before giving the recommendation of using a particular strain as a part of starter cultures is proposed. Ideally a strain that is intended to be used in a starter culture may not possess a single virulence factor and should be sensitive to relevant clinical antibiotics (Doming et al., 2003; Franz et al., 2003). On the basis of all the characteristics of enterococci, when it comes to the significance of their application, it is believed that they have some advantages, but also disadvantages.

Thanks to the great diversity of enterococci strains, possessing different functional characteristics, it is necessary to analyze each strain individually, before any use in food or medicine.

Based on these facts, it can be concluded that enterococci isolated from foods of animal origin must be seen with particular caution, since they are reservoirs of genes for antibiotic resistance and virulence genes, while at the same time produce biogenic amines in protein-rich foods.

Only the isolates that have a low proportion of tested factors should be considered as potential candidates for starter cultures.

ACKNOWLEDGMENTS

We would like to thank our collaborators Djordje Fira, Ivana Strahinić, Jelena Begović, Jelena Lozo, Branko Jovćić, Milica Nikolić, Jovanka Lukić, Brankica Filipić, Gordana Uzelac, and Marija Miljković for the work in the field of molecular genetics and the LAB application. This chapter would be impossible to write without their results. The authors are grateful to Nathaniel Aaron Sprinkle, a native English editor for the proofreading of the manuscript.

This work was founded by the Ministry of Education, Science and Technological Development, Republic of Serbia, grant OI 173019.

REFERENCES

Ananou, S., Garriga, M., Jofré, A., Aymerich, T., Gálvez, A., Maqueda, M., Martínez-Bueno, M., and Valdivia, E. 2010. Combined effect of enterocin AS-48 and high hydrostatic pressure to control food-borne pathogens inoculated in low acid fermented sausages. *Meat Sci.* 84: 594-600.

Andrighetto, C., Knijff, E., Lombardi, A., Torriani, S., Vancanneyt, M., Kersters, K., Swings, J., and Dellaglio, F. 2001. Phenotypic and genetic diversity of enterococci isolated from Italian cheeses. *J. Dairy Res.* 68: 303–316.

Asteri, I. A., Robertson, N., Kagkli, D. M., Andrewes, P., Nychas, G., Coolbear, T., Holland, R., Crow, V., and Tsakalidou, E. 2009. Technolo-gical and flavour potential of cultures isolated from traditional Greek cheese-A pool of novel species and starters. *Int. Dairy J.* 19: 595-604.

Canzek Majhenic, A., Rogelj, I. and Perko, B. 2005. Enterococci from Tolminc cheese: Population structure, antibiotic susceptibility and incidence of virulence determinants. *Int. J. Food Microbiol.* 102: 239–244.

Centeno, J. A., Menéndez, S. and Rodríguez-Otero, J. L. 1996. Main microbial flora present as natural starters in Cebreiro raw cow's milk cheese (North-west Spain). *Int. J. Food Microbiol.* 33: 307–313.

Clare, I., Konstabel, C., Badstubner, D., Werner, G., and Witte, W. 2003. Occurrence and spread of antibiotic resistances in *Enterococcus faecium*. *Int. J. Food Microbiol.* 88: 269-290.

Cleveland, J., Montville, T. J., Nes, I. F., and Chikindas, M. L. 2001. Bacterio-cins: safe, natural antimicrobials for food preservation. *Int. J. Food Micro-biol.* 71: 1–20.

De Vuyst, L., Foulquié, M. R. and Revets, H. 2003. Screening for enterocins and detection of hemolysin and vancomycin resistance in enterococci of different origins. *Int. J. Food Microbiol.* 84: 299–318.

Deegan, L. H., Cotter, P. D., Hill, C., and Ross, P. 2006. Bacteriocins: biological tools for bio-preservation and shelf-life extension. *Int. Dairy J.* 16: 1058–1071.

Domig, K. J., Mayer, H. K. and Kneifel, W. 2003. Methods used for the isolation, enumeration, characterisation and identification of *Enterococcus* spp.: 1. Media for isolation and enumeration. *Int. J. Food Microbiol.* 88: 147–164.

Du Toit, M., Franz, C. M. A. P., Dicks, L. M. T., and Holzapfel, W. H. 2000. Preliminary characterization of bacteriocins produced by *Enterococcus faecium* and *Enterococcus faecalis* isolated from pig faeces. *J. Appl. Microbiol.* 88: 482–494.

Eaton, J. T. and Gasson, J. M. 2001. Molecular screening of *Enterococcus* virulence determinants and potential for genetic exchange between food and medical isolates. *Appl. Environ. Microbiol.* 67: 1628–1635.

Fira, D., Kojic, M., Strahinic, I., Arsenijevic, S., Banina, A., and Topisirovic, L. 2000. Natural isolate *Enterococcus faecalis* BGPM3 produces an inducible extracellular proteinase. *Arch. Biol. Sci.* 52: 67–76.

Foulquie-Moreno, M. R., Sarantinopoulos, P., Tsakalidou, E., and De Vuyst, L. 2006. The role and application of enterococci in food and health. *Int. J. Food Microbiol.* 106: 1–24.

Franz, C. M. A. P., Holzapfel, W. H. and Stiles, M. E. 1999. Enterococci at the crossroads of food safety? *Int. J. Food Microbiol.* 47: 1–24.

Franz, C. M. A. P., Muscholl-Silberhorn, N., Yousif, M. K., Vancanneyt, M., Swings, J., and Holzapfel, W. H. 2001. Incidence of virulence factors and antibiotic resistance among enterococci isolated from food. *Appl. Environ. Microbiol.* 67: 4385-4389.

Franz, C. M. A. P., Stiles, M. E., Schleifer, K. H., and Holzapfel, W. H. 2003. Enterococci in foods–a conundrum for food safety. *Int. J. Food Microbiol.* 88: 105–122.

Franz, C. M. A. P., van Belkum, M. J., Holzapfel, W. H., Abriouel, H., and Gálvez, A. 2007. Diversity of enterococcal bacteriocins and their grouping in a new classification scheme. *FEMS Microbiol. Rev.* 31: 293–310.

Gálvez, A., Abriouel, H., López, R. L., and Ben Omar, N. 2007. Bacteriocin-based strategies for food biopreservation. *Int. J. Food Microbiol.* 120: 51–70.

Gardini, F., Martuscelli, M., Caruso, M. C., Galgano, F., Crudele, M. A., Favati, F., Guerzoni, M. E., and Suzzi, G. 2001. Effects of pH, tempera-ture and NaCl concentration on growth kinetics, proteolytic activity and biogenic amine production of *Enterococcus faecalis*. *Int. J. Food Micro-biol.* 64: 105–117.

Giménez–Pereira, M. L. 2005. Enterococci in milk products. *PhD thesis*, Massey University Palmerston North, New Zealand.

Giraffa, G., 1995. Enterococcal bacteriocins: their potential use as anti–Listeria factors in dairy technology. *Food Microbiol.* 12: 291-299.

Giraffa, G. 2002. Enterococci from foods. *FEMS Microbiol. Rev.* 744: 1–9.

Giraffa, G. 2003. Functionality of enterococci in dairy products. *Int. J. Food Microbiol.* 88: 215–222.

Golic, N., Cadez, N., Terzic-Vidojevic, A., Suranská, H., Beganovic, J., Lozo, J., Kos, B., Suskovic, J., Raspor, P., and Topisirovic, L. 2013. Evaluation of lactic acid bacteria and yeast diversity in traditional white pickled and fresh soft cheeses from the mountain regions of Serbia and lowland regions of Croatia. *Int. J. Food Microbiol.* 166 (2) 294-300.

Huys, G., D'Haene, K., Collard, J. M., and Swings, J. 2004. Prevalence and molecular characterization of tetracycline resistance in *Enterococcus* isolates from food. *Appl. Environ. Microbiol.* 70: 1555–1562.

Jamet, E., Akary, E., Poisson, M. A., Chamba, J. F., Bertrand, X., and Serror, P. 2012. Prevalence and characterization of antibiotic resistant *Entero-coccus faecalis* in French cheeses. *Food Microbiol.* 31: 191-198.

Jokovic, N., Nikolic, M., Begovic, J., Jovcic, B., Savic, D., and Topisirovic, L. 2008. A survey of the lactic acid bacteria isolated from Serbian artisanal dairy product kajmak. *Int. J. Food Microbiol.* 127: 305-311.

Kayser, F. H. 2003. Safety aspects of enterococci from the medical point of view. *Int. J. Food Microbiol.* 88: 255–262.

Leroy, F., Foulquie Moreno, M. R., and De Vuyst, L. 2003. *Enterococcus faecium* RZS C5, an interesting bacteriocin producer to be used as a co–culture in food fermentation. *Int. J. Food Microbiol.* 88: 235–240.

Lopes, M. F. S., Simões, A. P., Tenreiro, R., Marques J. J. F., and Crespo, M. T. B. 2006. Activity and expression of a virulence factor, gelatinase, in dairy enterococci. *Int. J. Food Microbiol.* 112: 208–214.

Lowe, A. M., Lambert, P. A. and Smith, A. W. 1995. Cloning of an *Entero-coccus faecalis* endocarditis antigen: homology with adhesins from some oral streptococci. *Infect. Immun.* 63: 703–706.

Malani, P. N., Kauffman, C. A. and Zervos, M. J. 2002. Enterococcal disease, epidemiology, and treatment. In: Gilmore, M. S., Clewell, D. B., Courvalin, P., Dunny, G. M., Murray, B. E., Rice, L. B. (Eds.), *The Enterococci: Pathogenesis, Molecular Biology, and Antibiotic resistance.* ASM, Washington, D.C., pp. 385–408.

Mannu, L., Paba, A., Daga, E., Comunian, R., Zanetti, S., Dupre, I., and Sechi, L. A. 2003. Comparison of the incidence of virulence determinants and antibiotic resistance between *Enterococcus faecium* strains of dairy, animal and clinical origin. *Int. J. Food Microbiol.* 88: 291–304.

Morea, M., Baruzzi, F. and Cocconcelli, P. S. 1999. Molecular and physiolo-gical characterization of dominant bacterial populations in traditional Mozzarella cheese processing. *J. Appl. Microbiol.* 87: 574–582.

Mundy, L. M., Sahm, D. F. and Gilmore, M. 2000. Relationships between enterococcal virulence and antimicrobial resistance. *Clin. Microbiol. Rev.* 13: 513-522.

Murray, B. E. and Weinstock, G. M. 1999. Enterococci: New aspects of an old organism. *Proc. Assoc. Am. Physicians.* 111: 328-334.

Nieto-Arribas, P., Seseña, S., Poveda, J. M., Chicón, R., Cabezas, L., Palop, L. 2011. *Enterococcus* populations in artisanal Manchego cheese: Biodiver-sity, technological and safety aspects. *Food Microbiol.* 28: 891-899.

Ogier, J. C. and Serror, P. 2008. Safety assessment of dairy microorganisms: The *Enterococcus* genus. *Int. J. Food Microbiol.* 126: 291–301.

Peters, J., Mac, K., Wichmann-Schauer, H., Klein, G., and Ellerbroek, L. 2003. Species distribution and antibiotic resistance patterns of enterococci isolated from food of animal origin in Germany. *Int. J. Food Microbiol.* 88: 311-314.

Randazzo, C. L., Pitino, I., De Luca, S., Scifo, G. O., and Caggia, C. 2008. Effect of wild strains used as starter cultures and adjunct cultures on the volatile compounds of the Pecorino Siciliano cheese. *Int. J. Food Micro-biol.* 122: 269-278.

Ribeiro, T., Abrantes, M., Fatima Silva Lopes, M. F., and Barreto Crespo, M. T. 2007. Vancomycin-susceptible dairy and clinical enterococcal isolates carry *vanA* and *vanB* genes. *Int. J. Food Microbiol.* 113: 289–295.

Sarantinopoulos, P., Andrighetto, C., Georgalaki, M. D., Rea, M. C., Lombardi, A., Cogan, T. M., Kalantzopoulos, G., and Tsakalidou, E. 2001. Biochemical properties of enterococci relevant to their technological performance. *Int. Dairy J.* 11: 621-647.

Suzzi, G., Caruso, M., Gardini, F., Lombardi, A., Vannini, L., Guerzoni, M. E., Andrighetto, C., and Lanorte, M. T. 2000. A survey of the enterococci isolated from an artisanal Italian goat's cheese (Semicotto Caprino). *J. Appl. Microbiol.* 89: 267-274.

Tannock, G. W. and Cook, G. 2002. Enterococci as members of the intestinal microflora of humans. In: Gilmore, M. S., Clewell, D. B., Courvalin, P., Dunny, G. M., Murray, B. E., and Rice, L. B. (Eds.), *The enterococci: pathogenesis, molecular biology, and antibiotic resistance*. ASM, Washington, D. C., pp. 101–132.

Terzic-Vidojevic, A., Vukasinovic, M., Veljovic, K., Ostojic, M., and Topisirovic, L. 2007. Characterization of microflora in homemade semi-hard white Zlatar cheese. *Int. J. Food Microbiol.* 114: 36-42.

Terzic-Vidojevic, A., Mihajlovic, S., Uzelac, G., Veljovic, K., Tolinacki, M., Nikolic, M., Topisirovic, L., and Kojic, M. 2014. Characterization of lactic acid bacteria isolated from artisanal Travnik young cheeses, sweet creams and sweet kajmaks over four seasons. *Food Microbiol.* 39:27-38.

Teuber, M., Meile, L. and Schwarz, F. 1999. Acquired antibiotic resistance in lactic acid bacteria from food. *Antonie Van Leeuwenhoek*, 76: 115–137.

Topisirovic, L., Kojic, M., Fira, D., Golic, N., Strahinic, I., and Lozo, J. 2006. Potential of lactic acid bacteria isolated from specific natural niches in food production and preservation. *Int. J. Food Microbiol.* 112: 230–235.

Tsakalidou, M. E., Tsilibari, V., Georgalaki, M., and Kalantzopoulos, G. 1993. Esterolytic activities of *Enterococcus durans* and *Enterococcus faecium* strains isolated from Greek cheese. *Neth. Milk Dairy J.* 145–150.

Valenzuela, A. S., ben Omar, N., Abriouel, H., López, R. L., Veljovic, K., Cañamero, M. M., Kojic, M., Topisirovic, L., and Gálvez, A. 2009. Virulence factors, antibiotic resistance, and bacteriocins in enterococci from artisan foods of animal origin. *Food Control* 20: 381–385.

Veljovic, K., Terzic–Vidojevic, A., Vukasinovic, M., Strahinic, I., Begovic, J., Lozo, J., Ostojic, M., and Topisirovic, L. 2007. Preliminary characteriza-tion of lactic acid bacteria isolated from Zlatar cheese. *J. Appl. Microbiol.* 103: 2142–2152.

Veljovic, K., Fira, D., Terzic–Vidojevic, A., Abriouel, H., Galvez, A., and Topisirovic, L. 2009. Evaluation of antimicrobial and proteolytic activity of enterococci isolated from fermented products. *Eur. Food Res. Tech.* 230: 63–70.

Veljovic, K., Terzic-Vidojevic, A., Tolinacki, M., Kojic, M., and Topisirovic, L. 2013. Molecular analysis of enterolysin A and *entl* gene cluster from natural isolate *Enterococcus faecalis* BGPT1-10P. *Genetika.* 45: 479-492.

Viedma, P. M., Abriouel, H., ben Omar, N., López, R. L., and Gálvez, A. 2009. Antistaphylococcal effect of enterocin AS-48 in bakery ingredients of vegetable origin, alone and in combination with selected antimicrobials. *J. Food Sci.* 84: 384-389.

In: *Enterococcus faecalis*
Editor: Henry L. Mack

ISBN: 978-1-63321-049-3
© 2014 Nova Science Publishers, Inc.

Chapter 8

ENTEROCOCCUS FAECALIS IN DENTAL INFECTIONS: VIRULENCE FACTORS, MOLECULAR CHARACTERISTICS, ANTIBACTERIAL AND ANTI-INFECTIVE TECHNIQUES

Nurit Beyth[1], Ronit Poraduso-Cohen[2] and Ronen Hazan[3],

[1]Department of Prosthodontics, the Hebrew University-Hadassah,
School of Dental Medicine, Jerusalem, Israel
[2]Department of Infectious Diseases of Sourasky
Medical Center and Tel-Aviv University, Israel
[3]Institute of Dental Sciences, The Hebrew University-Hadassah,
School of Dental Medicine, Jerusalem, Israel

ABSTRACT

Enterococcus faecalis is a commensal bacterium inhabiting the gastro-intestinal tract of humans. Interestingly, although it is not clear whether *E. faecalis* is part of the oral cavity microbiome, it is frequently recovered from root canal infections. Specifically, it is the major pathogen found in persistent infections associated with root canal treatment failure. Moreover, *E. faecalis* is one of the leading multidrug resistant nosocomial pathogens, causing infective endocarditis, and participating in urinary tract, wound, and device-device-related infections. The present chapter discusses *E. faecalis* virulence factors contributing to its high prevalence in nosocomial infections and root canal post treatment disease, including its ability to compete with other microorganisms, its cell to cell communication, its ability to invade various tissues, resist nutritional deprivation, facilitate the adherence of host cells and extracellular matrix, produce an immunomodulatory effect and cause toxin-mediated damage. Antiseptic techniques, conventional as well as novel, to overcome *the* survival ability of *E. faecalis* as well as virulence factors, are discussed in detail.

---

* Tel.: +972-2-6776142; Fax: +972-2-6429683; E-mail: ronenh@ekmd.huji.ac.il.

A. INTRODUCTION

Enterococci species are Gram-positive facultative anaerobe cocci that occur singly, in pairs or as short chains. They play an important role in human and animal microbiomes as a commensal of the gastro-intestinal tract and to a lesser extent in the female urogenital tract and the oral cavity (Garsin and Lorenz, 2013; Gilmore et al., 2013; Koch et al., 2004; Mason et al., 2011). On the other hand, *enterococci* are also among the leading multidrug resistant pathogens in hospital (nosocomial) diseases and as such have been of interest to researchers since the 1970s (Jett et al., 1994). *Enterococci* are a major cause of worldwide bacteremia, endocarditis (Dahl and Bruun, 2013), bacterial meningitis, penetrate the dentinal tubules (Stuart et al., 2006; Zehnder and Guggenheim, 2009b) urinary tract, wounds, and device-device related infections (Sava et al., 2010) among other infections. Thus, *enterococci,* and in particular *E. faecalis*, are considered one of the biggest challenges faced by medicine today (Eliopoulos, 2009). Moreover, *enterococci* are also of regulatory and industrial interest as they are used in food production, probiotic products and for tracking fecal contamination (Gilmore et al., 2013).

Although *enterococci* are commensal organisms situated in the intestinal and vaginal tracts and the oral cavity, it is recognized that the nosocomial *enterococci* infections are not caused by the patient's own flora, but rather originate in the hands of health care workers, clinical instruments, or other patients (Koch et al., 2004). *Enterococci* are also able to survive, multiply, and translocate from contaminated root canals to the draining submandibular lymph node (Ribeiro Sobrinho et al., 2001; Sobrinho et al., 1998).

Over the last few years, enterococci have received increasing attention because of the rapid spread of resistant *enterococci* to multiple antimicrobial drugs, specifically vancomycin. This may be the explanation for *enterococcal* dominance in nosocomial infections, which can be life-threatening in patients with serious infections or in immune-compromised patients. Particularly in dentistry enterococci pose an increasing problem, with therapy-resistant cases observed in endodontics (Kayaoglu and Orstavik, 2004; Zehnder and Guggenheim, 2009a).

Most enterococci infections are caused by *Enterococcus faecalis*, which is more likely to express virulence traits (Huycke et al., 1998; Rathnayake et al., 2012). In the 1950's Williams *et al.* detected enterococci in the oral cavity, although his findings were inconsistent (Williams et al., 1950). More recently, it has been stated that *E. faecalis* is a normal inhabitant of the oral cavity, although it is rarely detected in healthy mouths (Sedgley et al., 2004). Its prevalence was found to be increased in oral samples collected from patients undergoing endodontic treatment when compared with those with no endodontic history (Sedgley et al., 2004).

B. *E. FAECALIS* VIRULENCE FACTORS

The robust pathogenicity of *E. faecalis* is the result of an exceptional arsenal of virulence and adhesion factors (Giridhara Upadhyaya et al., 2009; Sava et al., 2010). Most of the knowledge of *E. faecalis* virulence factors was gained from gastro and endocarditis models. Nevertheless it is highly relevant also to endodonic infections (Kayaoglu and Orstavik, 2004), where biofilm is observed even in medicated root canals (Distel et al., 2002).

E. faecalis major secreted virulence factors include:

i. Gelatinase. The extracellular zinc metallo-endopeptidase gelatin- hydrolyzing protein gelatinase of *E. faecalis* is encoded by the gelE gene. In addition it also plays a role as a protease of many host cellular components. These include haemoglobin, casein, collagen and many small peptides (Kreft et al., 1992). This wide proteolytic activity is manifested in its importance in endocarditis infections (Gutschik et al., 1979). Downstream from gelE and co-transcribed with it is located sprE, encoding for a serine protease which is also annotated as a possible secreted protease (Lopes Mde et al., 2006).

With regard to endodonic infections, gelatinase activity was found in 70% of *E. faecalis* isolates from infected root canals (Sedgley et al., 2005), indicating its significance for bacterial survival in the host oral environment. In addition, high levels of gelatinase were observed in gingival biopsies (Soell et al., 2002), oral rinses, crevicular fluid and whole saliva (Makela et al., 1994) of periodontitis patients. Last, gelatinase inhibitors reduced bone resorption in a rat periodontal model (Ramamurthy et al., 2002).

ii. Secreted reactive oxygen species (ROS). *E. faecalis* uses ROS molecules superoxide and hydrogen peroxide as virulence factors against host and other microorganism competitors (Huycke et al., 1996). These extracellular ROS secreted by *E. faecalis* are significant for its in vivo survival in hosts (Huycke and Gilmore, 1997) as they damage the host cells' DNA (Huycke et al., 2002) and cause genome instability (Wang and Huycke, 2007) that possibly promote tumorigenesis and cancer (Strickertsson et al., 2013; Wang et al., 2008).

iii. Hemolysin and Cytolysin. *E. faecalis* harbors efficient cytolytic proteins that contribute to its virulence in many animal models and human infections by lysing erythrocytes (Chow et al., 1993; Ike et al., 1987). Expression of the hemolytic functions of *E. faecalis* is regulated by a quorum-sensing mechanism carried out by the cylR12 two-component system (Haas et al., 2002). Interestingly, it seems that hemolysin is not relevant for endodonic *E. faecalis* as none of 33 oral isolated strains harbor it (Sedgley et al., 2005).

In addition to its secreted factors, *E. faecalis* contains an arsenal of adhesion factors that allow it to attach to host cells and extracellular matrix:

i. Enterococcal surface protein (ESP). ESP also known as extracellular surface protein is encoded by the esp gene and is located in the cell wall of *E. faecalis* (Shankar et al., 1999). ESP was found to be involved in urinary tract infections (Leendertse et al., 2009), perhaps by inducing the adherence of *E. faecalis* to the bladder epithelium via uroplakin and mucin (Shankar et al., 2001). ESP was shown also to promote biofilm formation (Tendolkar et al., 2004; Toledo-Arana et al., 2001), although this observation is controversial (Kristich et al., 2004). Lastly, a correlation was found between the presence of ESP and clinical versus commensal isolations of *E. faecalis* (Shankar et al., 1999) and in 61% of endodonic isolates (Sedgley et al., 2005).

ii. Aggregation Substance (AS). *E. faecalis* undergoes horizontal gene transfer in which clinging cells aggregate and DNA forms, such as plasmids, are exchanged between cells via conjugation. This process is regulated by a secreted pheromone produced and sensed by the bacteria cell–cell communication system (Clewell, 1993). Cell aggregation is mediated by surface proteins termed aggregation substance that are induced by a secretion signal molecule. Interestingly, in vivo when *E. faecalis* is present in the host, AS expression is induced by its pheromones and perhaps also by host factors (Hirt et al., 2002). Furthermore, this induction was found to be tightly associated with an increase in pathogenicity. AS plays a role in the adherence to multiple host cells, including macrophages (Sussmuth et al., 2000), leukocytes (Rakita et al., 1999; Vanek et al., 1999), renal tubular cells (Kreft et al., 1992) and extracellular matrixes (Rozdzinski et al., 2001), all manifested in virulence as, for example, was shown for endocarditis (Chuang et al., 2009).

iii. Microbial surface components recognizing adhesive matrix molecules (MSCRAMM). In addition to ESP and AS, *E. faecalis* uses an array of surface proteins termed MSCRAMM to bind to the extracellular matrix of the host. Of the putative 17 MSCRAMM annotated in the *E. faecalis* genome it was found that several of them indeed function as adhesins which play a role in infections. The best studied MSCRAMM is Ace, a collagen (Nallapareddy et al., 2000a; Rich et al., 1999) dentin (Kowalski et al., 2006) and laminin (Nallapareddy et al., 2000a) a binding protein which was identified based on its homology to staphylococcal MSCRAMM protein Cha (Rich et al., 1999). In a study performed in the School of Dentistry, University of Michigan, Ace was found in all 33 endodontic enterococcal isolates (Sedgley et al., 2005).

The observation that 90% of patients with *E. faecalis*-related endocarditis had anti-Ace antibodies (Nallapareddy et al., 2000b) indicates that Ace is expressed in vivo. Furthermore, immunization with Ace or antibodies against it reduced the susceptibility of rats to *E. faecalis*-induced endocarditis (Singh et al., 2010).

Ace is of particular importance with respect to endodontic infections, as it binds to dentin via collagen type I (Hubble et al., 2003), one of the main organic components of dentin. In addition to Ace, three more MSCRAMM were found in *E. faecalis* and termed Fss1 – Fss3, which bind to fibrinogen polypeptide chains (Sillanpaa et al., 2009).

iv. Capsular and Cell Wall Carbohydrate (CCWC). One promising target for antibody therapy against *E. faecalis* are the polysaccharides and carbohydrates (Huebner et al., 2000) present on the cell surface as part of the bacterial cell wall. The most studied antigen among *E. faecalis* CCWC is the enterococcocal polysaccharide antigen (EPA) (Teng et al., 2002), which is abundant among enterococcal strains and of importance in biofilm formation, replication, lysis and virulence in mouse models of peritonitis (Teng et al., 2009) and urinary tract infection (Singh et al., 2009) models.

Additionally, it seems that CCWC are important for the pathogenicity of *E. faecalis* as, for example mutants in the cps operon, encoding for the biosynthesis of CCWC, were more susceptible to phagocytic killing in vitro and compromised in their ability to persist in regional lymph nodes *in vivo* (Hancock and Gilmore, 2002). Another linkage between CCWC and virulence arises from the observations that they play an important role in biofilm formation and resistance to antimicrobial peptides (Fabretti et al., 2006).

v. Glycolipids. Additional surface binding components of E. faecalis to host cells and extracellular matrix are its cell wall glycolipids. Glycolipids were found to be important for biofilm formation and for prolonged E. faecalis bacteremia (Theilacker et al., 2009). The glycolipid diglucosyldiacylglycerol (DGlcDAG) was found to be involved in the binding of the bacteria to heparin and heparan sulfate of colonic epithelial cells and extracellular matrix (Sava et al., 2009). Glycolipid mutants survive for a shorter period in the blood stream of infected mice, suggesting that this binding is important for the survival and virulence of the bacteria (Hufnagel et al., 2004; Theilacker et al., 2009).

vi. Pili. Last but not least among *E. faecalis* adhesion factors are the pili encoded by two pilin gene clusters (PGC): endocarditis and biofilmassociated pili (ebp) (Singh et al., 2007) and biofilm enhancer in enterococci (bee) (Tendolkar et al., 2006). Pili were found to be necessary for endocarditis infections by *E. faecalis* (Budzik and Schneewind, 2006;

Nallapareddy et al., 2006; Nielsen et al., 2013), biofilm formation (Nallapareddy et al., 2006; Nielsen et al., 2013) and urinary tract infections (Sillanpaa et al., 2013).

C. ANTIBIOTIC RESISTANCE OF *E. FAECALIS*

Enterococci come well equipped with a wide-ranging arsenal of intrinsic antibiotic resistance; they are also capable of acquiring new resistance genes and/or mutations. Nevertheless, until recently most enterococcal infections, could be treated with penicillin, ampicillin, or vancomycin with or without an aminoglycoside. Unfortunately, enterococci have now acquired resistance to these and many other agents as a result of mutations or the acquisition of new genes. The acquisition of resistance occurs by conjugation, using pheromone-responsive plasmids and their high frequency transfer between *E. faecalis* isolates (Dunny et al., 1995), conjugative plasmids with a broad host range or conjugative transposons with the potential to carry multiple antibiotic- resistant genes (Ehrenfeld et al., 1986). Much of the acquired resistance likely emerged among enterococci colonizing humans or animals that were given antibiotics for reasons other than enterococcal infections (Arias and Murray, 2012), (Hollenbeck and Rice, 2012).

ı. β - lactams and cephalosporines Problems in the treatment of enterococcal infections were noticed as early as the 1950s with the observation that enterococcal endocarditis was not cured with penicillin nearly as often as streptococcal endocarditis (Geraci and Martin, 1954). The reason for the poorer response appears to be that β lactam antibiotics such as penicillins are not as bactericidal against enterococci as against most viridans streptococci. This phenomenon, which has been commonly described as tolerance, (lack of killing despite growth inhibition), is characteristic of many enterococcal strains and, can be elicited in some strains by intermittent exposure to the antibiotic (Lakey and Ptak, 1988).

Enterococci express low affinity penicillin binding proteins (PBPs), PBP5 in *E. faecium* and PBP4 in *E. faecalis*. As a result, even with the more active beta-lactams (eg, penicillin, ampicillin, piperacillin), it takes 10 to 1000 times more drug to inhibit an average *enterococcus* than an average *streptococcus*. *E. faecalis*, the more susceptible of the two predominant enterococcal species, is usually inhibited by 1 to 4 mcg/mL of ampicillin and 2 to 8 mcg/mL of penicillin; the comparable minimal inhibitory concentrations (MICs) for *E. faecium* are typically 8 to 32 mcg/Ml (Sifaoui et al., 2001).

Ampicillin resistance, which is rare in *E. faecalis*, occurs in about 90% of hospital--associated *E. faecium* isolates. The mechanism involved is production of PBP5, called PBP5-R, which has a low affinity for penicillin. Higher levels of resistance to beta-lactam antibiotics appear to require one or more of the following mechanisms: expression of PBP5-R, alterations in the PBP5 protein around the active site (Rice et al., 2004), increased expression of PBP5, and utilization of a beta-lactam insensitive transpeptidase for cell wall synthesis (Mainardi et al., 2002).

β-lactamase production commonly seen in *S. aureus*, is rarely observed in enterococci, although there were outbreak strains of *E. faecalis* producing β-lactamse in the 1990s (Murray, 1992). These strains are not a therapeutic challenge as they are inhibited by β-lactamase inhibitor combinations (such as ampicillin-sulbactam). However, they may pose a diagnostic challenge, as at the standard inoculum used for susceptibility testing, penicillinase-producing enterococci appear no more resistant to penicillin or ampicillin than a non-penicillinase producing organism. Nonetheless, these organisms display resistance at a high inoculum, when more enzyme is present.

Enterococci are also intrinsically resistant to cephalosporins, with a high MIC for the majority of these compounds (lower MICs are seen with newer cephalosporins such as ceftobiprole and ceftaroline). Intrinsic resistance to cephalosporins in *E. faecalis* has been associated with the presence of a putative serine/threonine kinase (designated IreK) (Kristich et al., 2007).

ii. Aminoglycosides Enterococci exhibit intrinsic resistance to low and moderate levels of these drugs. The uptake of these highly polar molecules can be enhanced by the addition of cell wall-active agents, including penicillins and vancomycin (Chow, 2000). In this setting, there is usually a marked increase in killing (more killing than with penicillin alone), and a bactericidal effect (defined as ≥1000 CFU/mL or a 99.9 percent decrease from the starting inoculum) is achieved by 24 hours. This synergistic effect presumably explains the clinical observation in the 1950s that a combination of penicillin with streptomycin was much more effective in the treatment of enterococcal endocarditis than penicillin alone (Geraci and Martin, 1954). This regimen subsequently became the standard of care.

A naturally occurring characteristic of *E. faecium* is higher MICs of tobramycin (MICs 64 to 1000 mcg/mL) and resistance to synergism with this aminoglycoside. This is due to the presence of an aminoglycoside-modifying enzyme, a tobramycin 6'-acetyltransferase (AAC(6')-Ii), which modifies tobramycin but not gentamicin (Costa et al., 1993). The enzyme eliminates synergism between cell wall active agents and tobramycin, kanamycin, netilmicin and sisomicin. Similarly, many enterococci possess the aph(3')-IIIa gene, which confers high-level resistance to kanamycin and abolishes the synergistic effect of amikacin. Therefore, with minor exceptions, gentamicin and streptomycin are the only two aminoglycosides that should be considered to achieve synergistic therapy when treating enterococcal infections.

One of the more serious of the acquired resistance factors of enterococci is high-level resistance to both streptomycin and gentamicin. The significance of high-level resistance to these aminoglycosides is that it eliminates the expected synergism between gentamicin or streptomycin and a cell wall active agent such as penicillin or vancomycin. Streptomycin resistance can be caused by a ribosomal mutation, which leads to very high levels of resistance, or by the acquisition of a streptomycin-modifying enzyme (an adenylyltransferase) (Eliopoulos et al., 1984). High level gentamycin resistance is most often attributable to the production of an enzyme with two functional domains: one with 2'-phosphotransferase activity; and one with 6'-acetyltransferase activity. The combined enzymatic activities result in high-level resistance and/or resistance to synergism to all commercially available aminoglycosides, except streptomycin (Mingeot-Leclercq et al., 1999).

iii. Glycopeptides Until 1986, when high level vancomycin resistance to glycopeptides in enterococci emerged (Uttley et al., 1988), glycopeptides were the usual alternative for infections caused by ampicillin-resistant enterococci, or for patients with severe penicillin allergy.

Vancomycin inhibits enterococci by binding to the D-alanyl-D-alanine (D-Ala-D-Ala) terminus of cell wall precursors, compromising the subsequent enzymatic steps in synthesis of the cell wall. Vancomycin resistance is common in nature among Gram-positive cocci.

Examples are *E. gallinarum* and *E. casselifavus*, which produce peptidoglycan precursors with decreased affinity for vancomycin (Reynolds et al., 1999).

Glycopeptide resistance involves two pathways. One is replacement of the terminal D-alanine of peptidoglycan with D-lactate, which produces high level resistance, or with D-serine, which produces low level resistance. The D-lactate replacement eliminates one of five hydrogen bonds that would be established between vancomycin and the D-Ala-D-Ala termini, resulting in an almost 1000-fold decrease in affinity (Arthur and Courvalin, 1993). Replacement of D-Ala by D-Ser also decreases the binding affinity, but to a lesser extent. The second pathway involves prevention of destruction of precursors ending in D-Ala-D-Ala by specific D-dipeptidases and carboxypeptidases (Werner et al., 2008).

iv. Trimethoprim-sulfamethoxazole and clindamycin Although enterococci may be susceptible in vitro to trimethoprim-sulfamethoxazole, they appear to be resistant *in vivo* due to their capacity to utilize preformed folic acid (Grayson et al., 1990). Other resistance that appears to be naturally occurring in *E. faecalis* includes low-level resistance to clindamycin due to the presence of the gene lsa which encodes a putative ATP-binding protein (Singh et al., 2002).

v. Linezolid Linezolid, an oxazlidinone, is bacteriostatic and inhibits protein synthesis by interfering with the placement of aminoacyl tRNA at the A site of the bacterial ribosome. Resistance is generally associated with the extensive use of linezolid, although resistant enterococci have been isolated from patients without previous exposure to the antibiotic. The most common mutation in enterococci interferes with the positioning of crucial nucleotides in the linezolid binding site (Leach et al., 2007; Wilson et al., 2008).

vi. Daptomycin Daptomycin is a cyclic semisynthetic lipopeptide antibiotic that is active against a wide variety of Gram-positive bacteria. It has potent bactericidal activity, mainly due to its ability to penetrate the bacterial cell wall and enter the cytoplasmic **membrane** in a calcium-dependent manner. Daptomycin appears to target the membrane (interacting with phosphatidylglycerol [PG], a cell-membrane phospholipid) with preference for the bacterial septum, causing alterations in cell division, cell membrane architecture and function. Resistance to daptomycin has been increasingly reported in enterococci, including isolates from patients who have never received the antibiotic (Munoz-Price et al., 2005).

vii. Quinupristin-dalfopristin This is a mixture of two chemically unrelated compounds from the streptogramin family that act synergistically. Quinupristin-dalfopristin inhibits protein synthesis in bacteria by interfering with different targets on rRNA 23S in the 50S ribosomal subunit. *E. faecalis* is innately resistant to quinupristin-dalfopristin owing to the presence of the *lsa* gene, which encodes a putative ATP-binding efflux protein. Resistance to quinupristin-dalfopristin in *E. faecium* can occur by enzyme modification, hydrolysis, active transport, and target modification (Hershberger et al., 2004).

Treatment recommendations. Enterococci are relatively resistant to the killing effects of cell wall active agents (penicillin, ampicillin and vancomycin) and are impermeable to aminoglycosides. Therefore, a combined regimen of two agents, a cell wall-active agent with a synergistically active aminoglycoside or ampicillin with ceftriaxone,is required for optimal cure rates of invasive infections, such as endocarditis, likely meningitis, and certain cases of bacteremia. Ampicillin or penicillin are the preferred cell wall agents; vancomycin should be substituted only in the setting of beta-lactam resistance or hypersensitivity. Combinations of ampicillin or penicillin with gentamicin or streptomycin are preferable to vancomycin-aminoglycoside combinations as the latter pose a greater risk of nephrotoxicity. The combination of ampicillin and ceftriaxone has been associated with clinical cure rates equivalent to that of ampicillin plus gentamicin for *E. faecalis* endocarditis (Fernandez-Hidalgo et al., 2013).

The optimal approach for treatment of enterococcal infection due to vancomycin-resistant *E. faecium* (VRE) is uncertain. Although two agents (linezolid and quinupristin/dalfopristin) have been approved for use for infections caused by VRE, the utility of these agents for serious infections like endocarditis is uncertain.

Vancomycin-resistant *E. faecium* isolates often have concurrent high-level resistance to beta lactams and aminoglycosides. In contrast, vancomycin-resistant *E. faecalis* is usually susceptible to beta lactams, as is *E. gallinarum* and *E. casseliflavus* (which are intrinsically vancomycin-resistant). The newer agents linezolid, daptomycin, and tigecycline are active against both *E. faecalis* and *E. faecium*, whereas quinupristin-dalfopristin is active against *E. faecium* only.

Enterococci isolated from root canals are mainly *E, faecalis* and rarely *E. faecium*. Enterococcal isolates from the oral cavity resistant to ampicillin and clindamycin, commonly used in oral infections, have been described but vancomycin-resistant strains were not found (Dahlen et al., 2012), (Dahlen et al., 2000).

Summary. Because enterococci have shown the ability to develop resistance to essentially every antibiotic used against them, novel approaches are needed.

Studies should focus on ways of preventing enterococci from colonizing the gastrointestinal tract, understanding mechanisms of resistance to newer anti-enterococcal antibiotics such as daptomycin. Additional novel strategies, perhaps phage-based or combined therapeutic-immunological approaches, should also be considered.

D. *E. FAECALIS* QUORUM SENSING (QS) AND ITS RELATION TO VIRULENCE

Like most other studied bacteria and pathogens, *E. faecalis* contains cell-cell communication systems, also termed quorum sensing, which are based on secreted chemicals and pheromones that coordinate the activity of the whole culture to act as one organism (Carniol and Gilmore, 2004).

QS systems often regulate virulence factors in bacteria and *E. faecalis* is no exception. The best studied *E. faecalis* quorum sensing system is *the faecalis sensor regulator (Fsr)*, a paralog of the well-studied Agr system in *S. aureus* (Qin et al., 2000). *fsrC* and *fsrD* encode the synthesis of gelatinase biosynthesis-activating pheromone (GBAP), an auto-inducing cyclic peptide that is further processed and exported by FsrB. The response regulator FsrA senses GBAP and induces the expression of the *fsrBDC* in a positive autoregulatory circuit (Figure 1). Fsr regulates various virulence factors, including gelatinase and the putative serine protease SprE (Sifri et al., 2002). Via this regulation Fsr affects also other surface components, such as MSCRAMM Ace that significantly improve the adherence of *E. faecalis* to collagen (Pinkston et al., 2011).

A comprehensive study of the expression of Fsr- and GBAP-induced genes, performed in a drosophila infectious model demonstrated that in addition to *gel* and *sprE*, the Fsr regulon contains three more directly regulated genes: ef1097, ef1351 and ef1352. In addition, the expression of twelve other genes was found to be indirectly affected by the presence of GBAP via Fsr, among them the penicillin-related cell wall protein *lrgAB* and the two component system *lytRS* (Teixeira et al., 2013).

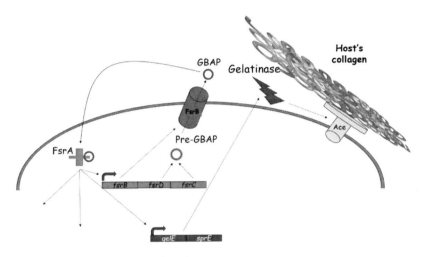

Figure 1. A model of the Fsr regulon. The cyclic peptide signal molecule GBAP is synthesized by FsrB and FsrC. Pre-GBPA is processed and secreted via FsrB. In the presence of its ligand, GBPA, the transcriptional regulator FsrA induces the expression of the fsr system in a positive loop as well as other virulence-related genes. One of these is *gelE*, expressing the virulence factor that plays a role both as a protease of host components and as an inducer of MSCRAMM Ace that attach the bacteria to the host' cells and extracellular matrix via collagen.

In support of the notion that the Fsr system is tightly involved in *E. faecalis* virulence is the observation that Fsr contributed to the pathogenicity of *E. faecalis* in many different infection models including: *C. elegans*, mice (Sifri et al., 2002), a rabbit model of endophthalmitis (Engelbert et al., 2004), rat endocarditis (Singh et al., 2005), *Arabidopsis thaliana* (Jha et al., 2005), insects (Gaspar et al., 2009; Teixeira et al., 2013) and recently in zebrafish, where Fsr was found to be responsible for tissue damage during *E. faecalis* infections (Prajsnar et al., 2013).

In addition to Fsr, *E. faecalis* harbors an auto-inducer, 2 (AI-2), a multi species quorum sensing system (Shao et al., 2012). In *E. faecalis* V583, AI-2 was shown to be involved in metabolism, translation, energy production, cell wall biogenesis and biofilm formation (Shao et al., 2012).

Aside from the Fsr and AI-2 systems, the *E. faecalis* genome also encodes for autoinducer 3 (AI-3) sensing system homologs. Recently, AI-3 was described as a signal and host epinephrine sensing system that controls many pathways in bacterial cells, including virulence, and is present in many species of bacteria (Karavolos et al., 2013). Nevertheless, to the best of our knowledge the impact of AI-3 on *E. faecalis* virulence has not been examined.

E. PREVALENCE OF *E. FAECALIS* IN ROOT CANAL INFECTIONS

Endodontic infections have a polymicrobial nature, with obligate anaerobic bacteria conspicuously dominating the microbiota in primary infections. However, the different niches in the root canal microenvironment with their limiting factors may favor and support the survival of one species relative to others. *E. faecalis* is commonly related to various forms of periradicular disease, including primary endodontic infections and persistent infections (Rocas et al., 2004).

Interestingly, *E. faecalis* has been frequently recovered from root canals undergoing retreatment, in cases of failed endodontic therapy and from canals with persistent infections associated with asymptomatic chronic periradicular lesions. It was less commonly isolated from teeth with acute periradicular periodontitis or acute periradicular abscesses (Love, 2001).

The reported prevalence of *E. faecalis* in primary endodontic infections ranges from 4 to 40%, whereas its prevalence is much higher post root canal treatment (up to 90%) (Baumgartner et al., 2004; Fouad et al., 2005; Hancock et al., 2001; Molander et al., 1998; Sundqvist et al., 1998).

Table 1. Comparison of PCR and culturing techniques in determining *E. faecalis* prevalence in failed root canals

E. faecalis Prevalence	Method	Source
65%	PCR	(Poptani et al., 2013)
38%	PCR	(Wang et al., 2012)
41%	PCR	(Zhu et al., 2010)
82%	PCR	(Zoletti et al., 2006)
67%	PCR	(Rocas et al., 2004)
32%	Culture	(Gomes et al., 2004)
77%	PCR	(Siqueira and Rocas, 2004)
46%	Culture	(Pinheiro et al., 2003)
33%	Culture	(Hancock et al., 2001)
64%	Culture	(Peciuliene et al., 2001)
38%	Culture	(Sundqvist et al., 1998)
47%	Culture	(Molander et al., 1998)

Failure of root canal treatment is often caused by the persistence of microorganisms in the root canal system after therapy or by recontamination of the root canal owing to bacterial microleakage. Actually, failed root canal treatments are nine times more likely to contain *E. faecalis* than are primary endodontic infections.

The wide range of the reported *E. faecalis* prevalence in the various studies may be attributed to different identification techniques. Studies using pure cultures recovered *E. faecalis* from root canals in ~35% of the cases of treated teeth with periradicular lesions (Pinheiro et al., 2003; Sundqvist et al., 1998).

Using the polymerase chain reaction (PCR) method for the detection of *E. faecalis*, most studies have shown a consistently higher prevalence of *E. faecalis* in failed root canal treatments when compared with culturing techniques (Table 1). PCR is advantageous over conventional culturing techniques as it is more sensitive, faster and may produce more accurate results.

An alternative approach claims that there is lack of evidence to support the concept that certain microorganisms of the microbial flora in root canal infections are more virulent than others, e.g. *E. faecalis*. Accordingly, because the commonly used isolation techniques are limited, the frequent reports of monocultures of *E. faecalis* in persistent root canal infections has raised suspicions whether this bacterium is the sole organism persisting in the root canals (Chavez de Paz, 2007).

F. Course of Disease and Severity

Endodontic infection is defined as the infection of the dental root canal system (Figure 2). Apical periodontitis is described as an infectious disease caused by microorganisms colonizing the root canal.

Although various chemical and physical factors can induce periradicular inflammation, evidence shows that microorganisms are the basis for the development and spread of different forms of apical periodontitis. In fact, a tooth with an infected non vital pulp is a reservoir of infection that is isolated from the patients' immune response and will eventually produce a periradicular inflammatory response. In some cases the microbes invade the periradicular tissues, resulting in the development of abscesses and cellulitis (Figure 2). The severity of the infection depends on the pathogenicity of the microbes and the resistance of the host.

Consequently, endodontic treatment aims to eradicate the infection and to prevent microorganisms from infecting or re-infecting the root and/or periradicular tissues. *E. faecalis* is often found in the root canal system of teeth with chronic apical periodontitis (Souto and Colombo, 2008; Wang et al., 2012) and even more so in post-treated root canal teeth compared with primary infections (Love, 2001; Sunde et al., 2002). It has been speculated that food-borne *E. faecalis* could enter the root canal system, and that the unique conditions of the root canal microenvironment support the survival of *E. faecalis* and the establishment of long-standing local infections (Kampfer et al., 2007). *E. faecalis* has many distinct features which make it an exceptional survivor in the root canal. It can persist in a poor nutritional environment, survive in the presence of various medications (e.g., calcium hydroxide) and irrigants (e.g., sodium hypochlorite), form biofilms in medicated canals, invade and metabolize fluids within the dentinal tubules and adhere to collagen, convert into a viable but non-cultivable state, acquire antibiotic resistance, survive in extreme environments with a low pH, high salinity and high temperatures, and endure prolonged periods of starvation and utilize tissue fluid that flows from the periodontal ligament (Narayanan and Vaishnavi, 2010).When this bacterium is present in small numbers, it is easily eliminated; but if it is present in large numbers, it is difficult to eradicate.

E. faecalis can gain entry into the root canal system during treatment, between appointments or after the treatment has been completed. Microorganisms may reach the root canal system through various routes, including: the dentinal tubules, the gingival sulcus that reaches the pulp chamber through the periodontal membrane, leakage through faulty restorations and through direct pulp exposure in cases of physical barrier breaks (Narayanan and Vaishnavi, 2010). As stated above, *E. faecalis* is seldom present in primary endodontic infections, but it is the principal microorganism in cases of post-endodontic treatment with apical periodontitis.

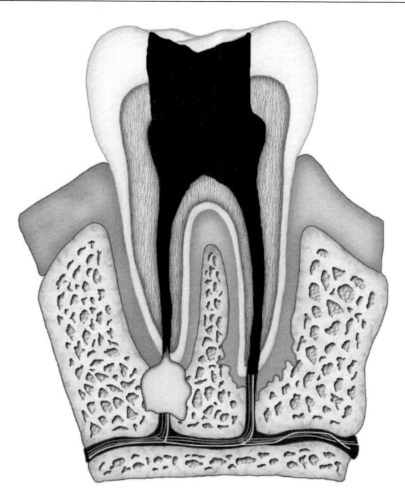

Figure 2. Pulp infection resulting in acute or chronic inflammatory lesions around the apex (bottom of the root), i.e. periapical periodontitis (also termed apical periodontitis or periradicular periodontitis). The infection may lead to abscesses in the soft tissue at the tip of the root canal system (left root). In some cases the swelling may spread to the surrounding tissues causing osteomyelitis or cellulitis (right root). The root canal system cavity is marked in black.

The purpose of endodontic treatment is to remove bacterial infection from the root canal system by biomechanical cleaning, shaping and disinfection. Once the bacteria are removed and the canals are properly shaped, the root canal is obturated and the tooth is sealed with a filling or crown. Obturation aims to prevent reinfection of the root canals. Proper obturation involves the use of materials and techniques capable of sealing the entire root canal system and providing a tight fluid seal from the apical portion of the canal. In most cases, the periradicular inflammation at the apical portion of the root will heal naturally. Nonetheless, adequate crown rehabilitation or crown restoration is required to prevent oral bacterial microleakage from the coronal portion of the tooth (Valadares et al., 2011). Treatment success is dependent both on the quality of the obturation and the final restoration. Ideally, at the end of the endodontic treatment, the root canals will be sterile, but given the complex anatomy of the root canal system, it is widely accepted that achieving this goal, using the available materials and techniques, is undoubtedly impossible. Root canals that were inadequately cleaned, shaped and obturated may have remains of infected debris, thus providing nutritional resources, and available leakage space for bacteria. Subsequently, when the opportunity arises, *E. faecalis* could act as a source of reinfection.

E. faecalis possesses certain features which make it an exceptional survivor in the root canal system. These microorganisms can survive in harsh conditions and overcome the challenges posed within the root canal system.

E. faecalis can persist in a poor nutrient environment. Microorganisms such as *E. faecalis* can survive within the small canals of apical ramifications or in the space between the root filling and canal wall. In fact, *E. faecalis* strains can survive over months in an environment where nutrients are scant and when commensality with other bacteria is reduced (Peciuliene et al., 2008). It endures extended periods of nutritional deficiency by invading and metabolizing fluids within the dentinal tubules or serum from the periodontal ligament (Figdor et al., 2003) and by adhering to the collagen fibers (Hubble et al., 2003; Love, 2001). If at the onset of starvation there is a high enough cell density, *E. faecalis* can withstand prolonged periods of starvation in a minimal metabolic state and thus it is capable of converting into a viable but non-cultivable state (Figdor et al., 2003). Starved cells have the capacity to recover upon addition of human serum. These microorganisms have been shown to, survive in extreme pH conditions (McHugh et al., 2004; Nakajo et al., 2006), high temperatures (Hartke et al., 1998), and even suppress the action of lymphocytes causing an altered host response (Lee et al., 2004). *E. faecalis* was reported to have a widespread genetic polymorphism that contributes to its survival and persistence ability (Sedgley et al., 2004). When forming a biofilm it was found to be highly resistant to antibodies, antimicrobials and phagocytosis (Distel et al., 2002). Moreover, it can be resistant to intracanal medicaments (e.g., calcium hydroxide) and irrigants (e.g., sodium hypochlorite) (Liu et al., 2010).

G. CURRENT TREATMENT MODALITIES AGAINST *E. FAECALIS* IN ROOT CANALS

E. faecalis can colonialize in the root canal system during treatment, between appointments, or post-treatment. Consequently, efforts are being directed towards finding an efficient way to eradicate *E. faecalis* from the root canal system. When the endodontic treatment is well-performed under proper aseptic conditions the success rate is generally high (85-90%) (Sjogren et al., 1990). Therefore, accepted clinical principles and meticulous treatment regimens should be followed to prevent *E. faecalis* infection during treatment or eliminate possible recontamination post-treatment. It has been suggested that larger size instrumentation of the root canal can eliminate *E. faecalis* contamination in the apical portion of the root canal (Card et al., 2002). Because the apical area is more difficult to disinfect by typical instrumentation regimens. Larger size instrumentation can eradicate bacteria from the canal more effectively by removing intratubular bacteria from the innermost dentin and by exposing the dentinal tubules to more efficient penetration of antimicrobial medicaments (Card et al., 2002).

i. Sodium hypochlorite is the most widely used irrigation solution in endodontic treatment. It has a broad spectrum of antimicrobial activity, rapidly kills vegetative and spore-forming bacteria, fungi, protozoa, and viruses (Siqueira et al., 2007). It was found to be effective against all *E. faecalis* growth phases including *E. faecalis* existing in a biofilm (Abdullah et al., 2005). Sodium hypochlorite oxidizes and hydrolyzes cell proteins, causes peptide link breakdown and, as a result, protein dissolution. In addition to its effective antibacterial effect, sodium hypochlorite dissolves necrotic tissue and organic components during root canal irrigation (Haapasalo et al., 2010). Various studies showed that 0.5-5.25% sodium hypochlorite irrigation is highly effective against *E. faecalis* (Berber et al., 2006; Siqueira et al., 1997; Vianna et al., 2004). Radcliffe et al. observed that *E. faecalis* was more resistant to sodium hypochlorite compared with yeast and *Candida albicans* (Radcliffe et al., 2004).

The efficacy of NaOCl has been confirmed in various clinical studies, which show a reduction in microbial count within the treated root canals. Nonetheless, root canals continue to harbor bacteria no matter what mechanical technique is used for preparation, the size of the apical preparation or the NaOCl concentration (Siqueira et al., 2007). In corroboration with *in vitro* results, when comparing the clinical persistence of *E. faecalis* with the persistence of Gram-negative rods and *Candida albicans* following NaOCl irrigation in re-treated root canals, *E. faecalis* shows the greatest persistence abilities (Peciuliene et al., 2001).

It has been speculated that the complex milieu and the variety of inorganic and organic compounds exert an inhibitory effect on the antibacterial effectiveness of irrigation medicaments. NaOCl dissolves organic matter in the root canal system. Although sodium hypochlorite has a higher capacity to dissolve organic tissue compared with other irrigants (Naenni et al., 2004; Okino et al., 2004), it has been shown that dentine powder has an evident inhibitory effect on sodium hypochlorites' antibacterial efficacy against *E. faecalis* (Haapasalo et al., 2010).

Various concentrations of sodium hypochlorite have been reported to be efficient root canal irrigants. Some studies suggested there is no significant difference in the antibacterial efficacy when using concentrations ranging from 0.5% to 5.25 % (Bystrom and Sundqvist, 1985; Siqueira et al., 2000). *In vitro* studies, particularly the ones testing the antibacterial efficacy using the agar diffusion test, have shown that the antibacterial effect depends on the NaOCl concentration (Siqueira et al., 1998; Yesilsoy et al., 1995). Others have found that even at low concentrations complete bacterial killing is achieved, although this occurs within 10-30 minutes at low NaOCl concentrations, whereas at higher concentrations it is a matter of seconds (Gomes et al., 2001). *Ex vivo* models and clinical studies have shown no significance differences in antibacterial efficacy between 0.5% and 5% NaOCl solutions (Bystrom and Sundqvist, 1985; Siqueira et al., 2000).

Despite all the advantages discussed above, the main disadvantage of NaOCl is that it can be toxic and cause tissue irritation, especially when used at high concentrations (Hauman and Love, 2003a, b).

ii. Calcium hydroxide is recognized as one of the most effective antimicrobial dressing materials available for endodontic treatment. Its advantages include the ability to stimulate the formation of calcified tissue, its ability to be bactericidal, and the capacity to denature protein, thus supporting the dissolution of pulpal remnants. Specifically, it is highly effective in endodontic cases where the tooth root is not fully developed, leading to sterile necrosis and subsequent calcification.

The calcium hydroxide mode of action is achieved through the ionic dissociation of Ca^{2+} and OH^- ions. Its antimicrobial effect involves damage to the cytoplasmic membrane, and protein and DNA denaturation in microbial cells. Calcium hydroxide pastes were reported effective in the elimination of *E. faecalis* in the dentinal tubules (Han et al., 2001). Moreover, the antimicrobial efficacy was found to be dependent on the viscosity of the paste used, i.e. the antimicrobial effect in less viscous pastes affects *E. faecalis* within the depth of the tubules (Behnen et al., 2001). The susceptibility of *E. faecalis* to calcium hydroxide depends on the bacterial cell phase. Cells in the exponential growth phase are the most sensitive and are killed within minutes, cells in the stationary phase are more resistant, but the most resistant are starved cells that cannot be totally eliminated (Portenier et al., 2005). Although it is recognized that calcium hydroxide has a wide range of antimicrobial activity, several studies have demonstrated that it is less effective against *E. faecalis* (Mohammadi and Dummer, 2011).

Survival of *E. faecalis* when exposed to calcium hydroxide appears to be related to a functioning proton pump that allows *E. faecalis* to withstand high pH levels (Evans et al., 2002). Moreover, only pH levels higher than 11.5 exert a bactericidal effect against *E. faecalis* (Bystrom and Sundqvist, 1985). It can be speculated that the ability *of E. faecalis* to invade dentinal tubules and its ability to form biofilm contribute to its non-susceptibility to calcium hydroxide.

Physicochemical properties of calcium hydroxide may limit its effectiveness in disinfecting the entire root canal system. To be effective as an antibacterial agent the ionic diffusion needs to exceed the buffering capacity of the dentine (Siqueira and Lopes, 1999). Interestingly, calcium hydroxide was found to be ineffective against *E. faecalis* in the presence of dentine powder (Haapasalo et al., 2000). Dentine has a buffering capacity because of the proton donors present in the inorganic hydroxyapatite. Nonetheless, organic materials present in the dentine may also reduce the antibacterial effect of calcium hydroxide against *E. faecalis* (Haapasalo et al., 2010). However, these are *in vitro* findings and should therefore be considered carefully as they have not been proved clinically.

In summary, calcium hydroxide is of great value in endodontic treatment, although its antimicrobial efficacy, specifically in retreatment cases, is controversial.

iii. Chlorhexidine (CHX) is a cationic bisbiguanide. At physiologic pH, chlorhexidine salts dissociate and release a positively charged chlorhexidine cation. It is efficacious against Gram-positive and Gram-negative organisms, facultative anaerobes, aerobes, and yeasts. The positive charge of the CHX cation interacts with the negatively charged bacterial cell wall, altering its permeability. At low concentrations chlorhexidine has a bacteriostatic effect; at high concentrations it causes membrane disruption, culminating in cell death. Despite the belief that CHX is biocompatible, it was reported to be toxic when in direct contact with human cells (Chang et al., 2001).

At high concentrations (2%), CHX reduces or completely eliminates *E. faecalis* from the root canal system, including the dentinal tubules (Basrani et al., 2002; Gomes et al., 2003). A 2% CHX rinse can remove *E. faecalis* from the superficial layers (up to 100 μm) of the dentinal tubules (Vahdaty et al., 1993). Moreover, it was shown to be highly efficient for up to two weeks (Gomes et al., 2003). At high concentrations a bactericidal effect is achieved owing to its ability to extract cytoplasmic contents and leading to bacterial death. Most comparative studies evaluating the efficacy of CHX and NaOCl against *E. faecalis* report a similar antibacterial effect (Gomes et al., 2001; Siqueira et al., 1998). In contrast, *in vitro* tests, using the agar diffusion test, have demonstrated the superiority of CHX over calcium hydroxide against *E. faecalis* (Ballal et al., 2007; Gomes et al., 2006; Neelakantan et al., 2007). In this test the inhibition zone of the tested material depends on its solubility in agar.

Therefore, small inhibition zones, when testing calcium hydroxide, can be attributed to its inability to diffuse in agar, and should not lead directly to the conclusion that it is less effective as an antibacterial agent.

CHX was shown to have an optimal antimicrobial effect in the pH range of 5.5-7.0. Thus, in combination with NaOCl the antibacterial effect of CHX will be reduced (Mohammadi and Dummer, 2011). It has been suggested that placement of CHX as a slow release medicament results in its deeper penetration into the dentinal tubules and a prolonged antibacterial effect against *E. faecalis* (Lin et al., 2003).

It has been considered that clinical protocols should include irrigation with NaOCl to dissolve organic components, and EDTA irrigation to eliminate the inorganic portion of the smear layer, thus allowing other irrigants (such as CHX) access to the dentinal tubules (Zehnder, 2006). Similarly to EDTA, CHX removes the smear layer but has little effect against *E. faecalis*. When using sodium benzoate in combination with citric acid there is a better chance for efficacious removal of this bacterium. Other irrigants that show an antibacterial effect against *E. faecalis* include a mixture of a tetracycline isomer detergent (MTAD) (Portenier et al., 2006), potassium iodide, ozonated water (Nagayoshi et al., 2004) and stannous fluoride (Mickel et al., 2003b). Various combinations of irrigants to eliminate *E. faecalis* have also been studied and have proved effective (Haapasalo et al., 2010). The antimicrobial efficacy against *E. faecalis* of numerous sealers has also been studied. Regrettably, most endodontic sealing materials (endodontic sealers) lack antibacterial properties; the few that were found to have an antibacterial effect lose this quality within a week. Sealers based on zinc-oxide eugenol, have been shown to be effective against *E. faecalis* (Mickel et al., 2003a). More recently it was found that affixation of polyethilenimine antimicrobial nanoparticles in an endodontic sealer produces long-lasting antimicrobial potency, providing an effective antimicrobial alternative (Beyth et al., 2013).

iv. Antibiotics are conventionally used in local application in endodontic treatment. System application is usually ineffective, specifically when dealing with necrotic pulpless teeth and periradicular tissues. Consequently, local application of antibiotics is more effective for delivery in endodontics.

Tetracyclines tetracycline-HCl, demeclocycline, minocycline and doxycycline, are broad spectrum antibiotics. Tetracyclines are bacteriostatic. In addition to their antibacterial effect they effectively removed the smear layer from instrumented root canal walls, and are used for irrigation of apical root-end cavities during periapical surgical procedures, and as an intracanal medicament (Mohammadi, 2009). *E. faecalis* isolated from root filled canals of teeth with periapical lesions was susceptible to tetracycline and doxycycline (Pinheiro et al., 2004). Tetracyclines attach to dentine and are released over time, thus creating an active reservoir of antibacterial agent. A conventionally used paste containing tetracycline is Ledermix paste. It consists of a tetracycline antibiotic, demeclocycline–HCl, and a corticosteroid, triamcinolone acetonide, in a polyethylene glycol base. MTAD (mixture of tetracycline, acid and detergent), Tetraclean (contains doxycycline) are relatively new root canal medicaments. All three have been shown to be effective in some degree against *E. faecalis*.

Clindamycin, neomycin and polymixin B sulphate are not ideally used in endodontic treatment, and are not recommended against *E. faecalis*. Unfortunately, *E. faecalis* displays intrinsic resistance to clindamycin (Singh et al., 2002).

CONCLUSION

E. faecalis is still a major problem in the clinic, in particular in dental health, due to its vast virulence factors and antibiotic resistance. Thus, the development of new combinations of adequate irrigants, medicaments and sealing materials with biocompatibility, together with novel antimicrobial and/or anti-infective treatments are critical for the success of root canal treatment.

REFERENCES

Abdullah, M., Ng, Y. L., Gulabivala, K., Moles, D. R., and Spratt, D. A. (2005). Susceptibilties of two *Enterococcus faecalis* phenotypes to root canal medications. *J. Endod.*, 31, 30-36.

Arias, C. A., and Murray, B. E. (2012). The rise of the *Enterococcus*: beyond vancomycin resistance. *Nat. Rev. Microbiol.*, 10, 266-278.

Arthur, M., and Courvalin, P. (1993). Genetics and mechanisms of glycopeptide resistance in enterococci. *Antimicrob. Agents Chemother.*, 37, 1563-1571.

Ballal, V., Kundabala, M., Acharya, S., and Ballal, M. (2007). Antimicrobial action of calcium hydroxide, chlorhexidine and their combination on endodontic pathogens. *Aust. Dent. J., 52*, 118-121.

Basrani, B., Santos, J. M., Tjaderhane, L., Grad, H., Gorduysus, O., Huang, J., Lawrence, H. P., and Friedman, S. (2002). Substantive antimicrobial activity in chlorhexidine-treated human root dentin. *Oral Surg. Oral Med. Oral Pathol. Oral Radiol. Endod.*, 94, 240-245.

Baumgartner, J. C., Siqueira, J. F., Jr., Xia, T., and Rocas, I. N. (2004). Geographical differences in bacteria detected in endodontic infections using polymerase chain reaction. *J. Endod.*, 30, 141-144.

Behnen, M. J., West, L. A., Liewehr, F. R., Buxton, T. B., and McPherson, J. C., 3rd (2001). Antimicrobial activity of several calcium hydroxide preparations in root canal dentin. *J. Endod.*, 27, 765-767.

Berber, V. B., Gomes, B. P., Sena, N. T., Vianna, M. E., Ferraz, C. C., Zaia, A. A., and Souza-Filho, F. J. (2006). Efficacy of various concentrations of NaOCl and instrumentation techniques in reducing *Enterococcus faecalis* within root canals and dentinal tubules. *Int. Endod. J.*, 39, 10-17.

Beyth, N., Kesler Shvero, D., Zaltsman, N., Houri-Haddad, Y., Abramovitz, I., Davidi, M. P., and Weiss, E. I. (2013). Rapid kill-novel endodontic sealer and *Enterococcus faecalis*. *PLoS One*, 8, e78586.

Budzik, J. M., and Schneewind, O. (2006). Pili prove pertinent to enterococcal endocarditis. *J. Clin. Invest.*, 116, 2582-2584.

Bystrom, A., and Sundqvist, G. (1985). The antibacterial action of sodium hypochlorite and EDTA in 60 cases of endodontic therapy. *Int. Endod. J.*, 18, 35-40.

Card, S. J., Sigurdsson, A., Orstavik, D., and Trope, M. (2002). The effectiveness of increased apical enlargement in reducing intracanal bacteria. *J. Endod.*, 28, 779-783.

Carniol, K., and Gilmore, M. S. (2004). Signal transduction, quorum-sensing, and extracellular protease activity in *Enterococcus faecalis* biofilm formation. *J. Bacteriol.*, 186, 8161-8163.

Chang, Y. C., Huang, F. M., Tai, K. W., and Chou, M. Y. (2001). The effect of sodium hypochlorite and chlorhexidine on cultured human periodontal ligament cells. *Oral Surg. Oral Med. Oral Pathol. Oral Radiol. Endod.*, 92, 446-450.

Chavez de Paz, L. E. (2007). Redefining the persistent infection in root canals: possible role of biofilm communities. *J. Endod.*, 33, 652-662.

Chow, J. W. (2000). Aminoglycoside resistance in enterococci. *Clin. Infect. Dis.*, 31, 586-589.

Chow, J. W., Thal, L. A., Perri, M. B., Vazquez, J. A., Donabedian, S. M., Clewell, D. B., and Zervos, M. J. (1993). Plasmid-associated hemolysin and aggregation substance production contribute to virulence in experimental enterococcal endocarditis. *Antimicrob. Agents Chemother.*, 37, 2474-2477.

Chuang, O. N., Schlievert, P. M., Wells, C. L., Manias, D. A., Tripp, T. J., and Dunny, G.M. (2009). Multiple functional domains of *Enterococcus faecalis* aggregation substance Asc10 contribute to endocarditis virulence. *Infect. Immun.*, 77, 539-548.

Clewell, D. B. (1993). Bacterial sex pheromone-induced plasmid transfer. *Cell,* 73, 9-12.

Costa, Y., Galimand, M., Leclercq, R., Duval, J., and Courvalin, P. (1993). Characterization of the chromosomal aac(6')-Ii gene specific for *Enterococcus faecium*. *Antimicrob. Agents Chemother.*, 37, 1896-1903.

Dahl, A., and Bruun, N. E. (2013). *Enterococcus faecalis* infective endocarditis: focus on clinical aspects. *Expert Rev. Cardiovasc. Ther.*, 11, 1247-1257.

Dahlen, G., Blomqvist, S., Almstahl, A., and Carlen, A. (2012). Virulence factors and antibiotic susceptibility in enterococci isolated from oral mucosal and deep infections. *J. Oral Microbiol.*, 4.

Dahlen, G., Samuelsson, W., Molander, A., and Reit, C. (2000). Identification and antimicrobial susceptibility of enterococci isolated from the root canal. *Oral Microbiol. Immunol.*, 15, 309-312.

Distel, J. W., Hatton, J. F., and Gillespie, M. J. (2002). Biofilm formation in medicated root canals. *J. Endod.*, 28, 689-693.

Dunny, G. M., Leonard, B. A., and Hedberg, P. J. (1995). Pheromone-inducible conjugation in *Enterococcus faecalis*: interbacterial and host-parasite chemical communication. *J. Bacteriol.*, 177, 871-876.

Ehrenfeld, E. E., Kessler, R. E., and Clewell, D. B. (1986). Identification of pheromone-induced surface proteins in *Streptococcus faecalis* and evidence of a role for lipoteichoic acid in formation of mating aggregates. *J. Bacteriol.*, 168, 6-12.

Eliopoulos, G. M. (2009). Microbiology of drugs for treating multiply drug-resistant Gram-positive bacteria. *J. Infect.*, 59 Suppl. 1, S17-24.

Eliopoulos, G. M., Farber, B. F., Murray, B. E., Wennersten, C., and Moellering, R. C., Jr. (1984). Ribosomal resistance of clinical *enterococcal* to *streptomycin* isolates. *Antimicrob. Agents Chemother.*, 25, 398-399.

Engelbert, M., Mylonakis, E., Ausubel, F.M., Calderwood, S.B., and Gilmore, M.S. (2004). Contribution of gelatinase, serine protease, and fsr to the pathogenesis of *Enterococcus faecalis* endophthalmitis. *Infect. Immun.*, 72, 3628-3633.

Evans, M., Davies, J. K., Sundqvist, G., and Figdor, D. (2002). Mechanisms involved in the resistance of *Enterococcus faecalis* to calcium hydroxide. *Int. Endod. J.*, 35, 221-228.

Fabretti, F., Theilacker, C., Baldassarri, L., Kaczynski, Z., Kropec, A., Holst, O., and Huebner, J. (2006). Alanine esters of enterococcal lipoteichoic acid play a role in biofilm formation and resistance to antimicrobial peptides. *Infect. Immun.*, 74, 4164-4171.

Fernandez-Hidalgo, N., Almirante, B., Gavalda, J., Gurgui, M., Pena, C., de Alarcon, A., Ruiz, J., Vilacosta, I., Montejo, M., Vallejo, N., *et al.* (2013). Ampicillin plus ceftriaxone is as effective as ampicillin plus gentamicin for treating *enterococcus faecalis* infective endocarditis. *Clin. Infect. Dis.*, 56, 1261-1268.

Figdor, D., Davies, J. K., and Sundqvist, G. (2003). Starvation survival, growth and recovery of *Enterococcus faecalis* in human serum. *Oral Microbiol. Immunol.*, 18, 234-239.

Fouad, A. F., Zerella, J., Barry, J., and Spangberg, L. S. (2005). Molecular detection of *Enterococcus* species in root canals of therapy-resistant endodontic infections. *Oral Surg. Oral Med. Oral Pathol. Oral Radiol. Endod.*, 99, 112-118.

Garsin, D. A., and Lorenz, M. C. (2013). Candida albicans and *Enterococcus faecalis* in the gut: Synergy in commensalism? *Gut microbes*, 4.

Gaspar, F., Teixeira, N., Rigottier-Gois, L., Marujo, P., Nielsen-LeRoux, C., Crespo, M. T., Lopes Mde, F., and Serror, P. (2009). Virulence of *Enterococcus faecalis* dairy strains in an insect model: the role of *fsrB* and *gelE*. *Microbiology*, 155, 3564-3571.

Geraci, J. E., and Martin, W. J. (1954). Antibiotic therapy of bacterial endocarditis. VI. Subacute enterococcal endocarditis; clinical, pathologic and therapeutic consideration of 33 cases. *Circulation*, 10, 173-194.

Gilmore, M. S., Lebreton, F., and van Schaik, W. (2013). Genomic transition of *Enterococci* from gut commensals to leading causes of multidrug-resistant hospital infection in the antibiotic era. *Curr. Opin. Microbiol.*, 16, 10-16.

Giridhara Upadhyaya, P. M., Ravikumar, K. L., and Umapathy, B. L. (2009). Review of virulence factors of *Enterococcus*: an emerging nosocomial pathogen. *Indian J. Med. Microbiol.*, 27, 301-305.

Gomes, B. P., Ferraz, C. C., Vianna, M. E., Berber, V. B., Teixeira, F. B., and Souza-Filho, F. J. (2001). In vitro antimicrobial activity of several concentrations of sodium hypochlorite and chlorhexidine gluconate in the elimination of *Enterococcus faecalis*. *Int. Endod. J.*, 34, 424-428.

Gomes, B. P., Pinheiro, E. T., Gade-Neto, C. R., Sousa, E. L., Ferraz, C. C., Zaia, A. A., Teixeira, F. B., and Souza-Filho, F. J. (2004). Microbiological examination of infected dental root canals. *Oral Microbiol. Immunol.*, 19, 71-76.

Gomes, B. P., Souza, S. F., Ferraz, C. C., Teixeira, F. B., Zaia, A. A., Valdrighi, L., and Souza-Filho, F. J. (2003). Effectiveness of 2% chlorhexidine gel and calcium hydroxide against *Enterococcus faecalis* in bovine root dentine *in vitro*. *Int. Endod. J.*, 36, 267-275.

Gomes, B. P., Vianna, M. E., Sena, N. T., Zaia, A. A., Ferraz, C. C., and de Souza Filho, F. J. (2006). *In vitro* evaluation of the antimicrobial activity of calcium hydroxide combined with chlorhexidine gel used as intracanal medicament. *Oral Surg. Oral Med. Oral Pathol. Oral Radiol. Endod.*, 102, 544-550.

Grayson, M. L., Thauvin-Eliopoulos, C., Eliopoulos, G. M., Yao, J. D., DeAngelis, D. V., Walton, L., Woolley, J. L., and Moellering, R. C., Jr. (1990). Failure of trimethoprim-sulfamethoxazole therapy in experimental enterococcal endocarditis. *Antimicrob. Agents Chemother.*, 34, 1792-1794.

Gutschik, E., Moller, S., and Christensen, N. (1979). Experimental endocarditis in rabbits. 3. Significance of the proteolytic capacity of the infecting strains of *Streptococcus faecalis*. *Acta Pathol. Microbiol. Scand. B*, 87, 353-362.

Haapasalo, H. K., Siren, E. K., Waltimo, T. M., Orstavik, D., and Haapasalo, M. P. (2000). Inactivation of local root canal medicaments by dentine: an *in vitro* study. *Int. Endod. J.*, 33, 126-131.

Haapasalo, M., Shen, Y., Qian, W., and Gao, Y. (2010). Irrigation in endodontics. *Dent. Clin. North Am.*, 54, 291-312.

Haas, W., Shepard, B. D., and Gilmore, M. S. (2002). Two-component regulator of *Enterococcus faecalis* cytolysin responds to quorum-sensing autoinduction. *Nature*, 415, 84-87.

Han, G. Y., Park, S. H., and Yoon, T. C. (2001). Antimicrobial activity of Ca(OH)2 containing pastes with *Enterococcus faecalis in vitro*. *J. Endod.*, 27, 328-332.

Hancock, H. H., 3rd, Sigurdsson, A., Trope, M., and Moiseiwitsch, J. (2001). Bacteria isolated after unsuccessful endodontic treatment in a North American population. *Oral Surg. Oral Med. Oral Pathol. Oral Radiol. Endod.*, 91, 579-586.

Hancock, L. E., and Gilmore, M. S. (2002). The capsular polysaccharide of *Enterococcus faecalis* and its relationship to other polysaccharides in the cell wall. *Proc. Natl. Acad. Sci. U S A*, 99, 1574-1579.

Hartke, A., Giard, J. C., Laplace, J. M., and Auffray, Y. (1998). Survival of *Enterococcus faecalis* in an oligotrophic microcosm: changes in morphology, development of general stress resistance, and analysis of protein synthesis. *Appl. Environ. Microbiol.*, 64, 4238-4245.

Hauman, C. H., and Love, R. M. (2003a). Biocompatibility of dental materials used in contemporary endodontic therapy: a review. Part 1. Intracanal drugs and substances. *Int. Endod. J.*, 36, 75-85.

Hauman, C. H., and Love, R. M. (2003b). Biocompatibility of dental materials used in contemporary endodontic therapy: a review. Part 2. Root-canal-filling materials. *Int. Endod. J.*, 36, 147-160.

Hershberger, E., Donabedian, S., Konstantinou, K., and Zervos, M. J. (2004). Quinupristin-dalfopristin resistance in gram-positive bacteria: mechanism of resistance and epidemiology. *Clin. Infect. Dis.*, 38, 92-98.

Hirt, H., Schlievert, P. M., and Dunny, G. M. (2002). *In vivo* induction of virulence and antibiotic resistance transfer in *Enterococcus faecalis* mediated by the sex pheromone-sensing system of pCF10. *Infect. Immun.*, 70, 716-723.

Hollenbeck, B. L., and Rice, L. B. (2012). Intrinsic and acquired resistance mechanisms in enterococcus. *Virulence*, 3, 421-433.

Hubble, T. S., Hatton, J. F., Nallapareddy, S. R., Murray, B. E., and Gillespie, M. J. (2003). Influence of *Enterococcus faecalis* proteases and the collagen-binding protein, Ace, on adhesion to dentin. *Oral Microbiol. Immunol.*, 18, 121-126.

Huebner, J., Quaas, A., Krueger, W. A., Goldmann, D. A., and Pier, G. B. (2000). Prophylactic and therapeutic efficacy of antibodies to a capsular polysaccharide shared among vancomycin-sensitive and -resistant enterococci. *Infect. Immun.*, 68, 4631-4636.

Hufnagel, M., Koch, S., Creti, R., Baldassarri, L., and Huebner, J. (2004). A putative sugar-binding transcriptional regulator in a novel gene locus in *Enterococcus faecalis* contributes to production of biofilm and prolonged bacteremia in mice. *J. Infect. Dis.*, 189, 420-430.

Huycke, M. M., Abrams, V., and Moore, D. R. (2002). *Enterococcus faecalis* produces extracellular superoxide and hydrogen peroxide that damages colonic epithelial cell DNA. *Carcinogenesis*, 23, 529-536.

Huycke, M. M., and Gilmore, M. S. (1997). In vivo survival of *Enterococcus faecalis* is enhanced by extracellular superoxide production. *Adv. Exp. Med. Biol.*, 418, 781-784.

Huycke, M. M., Joyce, W., and Wack, M. F. (1996). Augmented production of extracellular superoxide by blood isolates of *Enterococcus faecalis*. *J. Infect. Dis.*, 173, 743-746.

Huycke, M. M., Sahm, D. F., and Gilmore, M. S. (1998). Multiple-drug resistant enterococci: the nature of the problem and an agenda for the future. *Emerg. Infect. Dis.*, 4, 239-249.

Ike, Y., Hashimoto, H., and Clewell, D. B. (1987). High incidence of hemolysin production by *Enterococcus* (*Streptococcus*) *faecalis* strains associated with human parenteral infections. *J. Clin. Microbiol.*, 25, 1524-1528.

Jett, B. D., Huycke, M. M., and Gilmore, M. S. (1994). Virulence of enterococci. *Clin. Microbiol. Rev.*, 7, 462-478.

Jha, A. K., Bais, H. P., and Vivanco, J. M. (2005). *Enterococcus faecalis* mammalian virulence-related factors exhibit potent pathogenicity in the *Arabidopsis thaliana* plant model. *Infect. Immun.*, 73, 464-475.

Kampfer, J., Gohring, T. N., Attin, T., and Zehnder, M. (2007). Leakage of food-borne *Enterococcus faecalis* through temporary fillings in a simulated oral environment. *Int. Endod. J.*, 40, 471-477.

Karavolos, M. H., Winzer, K., Williams, P., and Khan, C. M. (2013). Pathogen espionage: multiple bacterial adrenergic sensors eavesdrop on host communication systems. *Mol. Microbiol.*, 87, 455-465.

Kayaoglu, G., and Orstavik, D. (2004). Virulence factors of *Enterococcus faecalis*: relationship to endodontic disease. *Crit. Rev. Oral Biol. Med.*, 15, 308-320.

Koch, S., Hufnagel, M., Theilacker, C., and Huebner, J. (2004). Enterococcal infections: host response, therapeutic, and prophylactic possibilities. *Vaccine*, 22, 822-830.

Kowalski, W. J., Kasper, E. L., Hatton, J. F., Murray, B. E., Nallapareddy, S. R., and Gillespie, M. J. (2006). *Enterococcus faecalis* adhesin, Ace, mediates attachment to particulate dentin. *J. Endod.*, 32, 634-637.

Kreft, B., Marre, R., Schramm, U., and Wirth, R. (1992). Aggregation substance of *Enterococcus faecalis* mediates adhesion to cultured renal tubular cells. *Infect. Immun.*, 60, 25-30.

Kristich, C. J., Li, Y. H., Cvitkovitch, D. G., and Dunny, G. M. (2004). Esp-independent biofilm formation by *Enterococcus faecalis*. *J. Bacteriol.*, 186, 154-163.

Kristich, C. J., Wells, C. L., and Dunny, G. M. (2007). A eukaryotic-type Ser/Thr kinase in *Enterococcus faecalis* mediates antimicrobial resistance and intestinal persistence. *Proc. Natl. Acad. Sci. U S A*, 104, 3508-3513.

Lakey, J. H., and Ptak, M. (1988). Fluorescence indicates a calcium-dependent interaction between the lipopeptide antibiotic LY146032 and phospholipid membranes. *Biochemistry*, 27, 4639-4645.

Leach, K. L., Swaney, S. M., Colca, J. R., McDonald, W. G., Blinn, J. R., Thomasco, L. M., Gadwood, R. C., Shinabarger, D., Xiong, L., and Mankin, A. S. (2007). The site of action of oxazolidinone antibiotics in living bacteria and in human mitochondria. *Mol. Cell,* 26, 393-402.

Lee, W., Lim, S., Son, H. H., and Bae, K. S. (2004). Sonicated extract of *Enterococcus faecalis* induces irreversible cell cycle arrest in phytohemagglutinin-activated human lymphocytes. *J. Endod.,* 30, 209-212.

Leendertse, M., Heikens, E., Wijnands, L. M., van Luit-Asbroek, M., Teske, G. J., Roelofs, J. J., Bonten, M. J., van der Poll, T., and Willems, R. J. (2009). Enterococcal surface protein transiently aggravates Enterococcus faecium-induced urinary tract infection in mice. *J. Infect. Dis.,* 200, 1162-1165.

Lin, Y. H., Mickel, A. K., and Chogle, S. (2003). Effectiveness of selected materials against *Enterococcus faecalis:* part 3. The antibacterial effect of calcium hydroxide and chlorhexidine on *Enterococcus faecalis. J. Endod.,* 29, 565-566.

Liu, H., Wei, X., Ling, J., Wang, W., and Huang, X. (2010). Biofilm formation capability of *Enterococcus faecalis* cells in starvation phase and its susceptibility to sodium hypochlorite. *J. Endod.,* 36, 630-635.

Lopes Mde, F., Simoes, A. P., Tenreiro, R., Marques, J. J., and Crespo, M. T. (2006). Activity and expression of a virulence factor, gelatinase, in dairy enterococci. *Int. J. Food Microbiol.,* 112, 208-214.

Love, R. M. (2001). *Enterococcus faecalis*--a mechanism for its role in endodontic failure. *Int. Endod. J.,* 34, 399-405.

Mainardi, J. L., Morel, V., Fourgeaud, M., Cremniter, J., Blanot, D., Legrand, R., Frehel, C., Arthur, M., Van Heijenoort, J., and Gutmann, L. (2002). Balance between two transpeptidation mechanisms determines the expression of beta-lactam resistance in *Enterococcus faecium. J. Biol. Chem.,* 277, 35801-35807.

Makela, M., Salo, T., Uitto, V. J., and Larjava, H. (1994). Matrix metalloproteinases (MMP-2 and MMP-9) of the oral cavity: cellular origin and relationship to periodontal status. *J. Dent. Res.,* 73, 1397-1406.

Mason, K. L., Stepien, T. A., Blum, J. E., Holt, J. F., Labbe, N. H., Rush, J. S., Raffa, K. F., and Handelsman, J. (2011). From commensal to pathogen: translocation of *Enterococcus faecalis* from the midgut to the hemocoel of Manduca sexta. *mBio,* 2, e00065-00011.

McHugh, C. P., Zhang, P., Michalek, S., and Eleazer, P. D. (2004). pH required to kill *Enterococcus faecalis in vitro. J. Endod.,* 30, 218-219.

Mickel, A. K., Nguyen, T. H., and Chogle, S. (2003a). Antimicrobial activity of endodontic sealers on *Enterococcus faecalis. J. Endod.,* 29, 257-258.

Mickel, A. K., Sharma, P., and Chogle, S. (2003b). Effectiveness of stannous fluoride and calcium hydroxide against *Enterococcus faecalis. J. Endod.,* 29, 259-260.

Mingeot-Leclercq, M. P., Glupczynski, Y., and Tulkens, P. M. (1999). Aminoglycosides: activity and resistance. *Antimicrob. Agents Chemother.,* 43, 727-737.

Mohammadi, Z. (2009). Antibiotics as intracanal medicaments: a review. *J. Calif. Dent .Assoc.,* 37, 98-108.

Mohammadi, Z., and Dummer, P. M. (2011). Properties and applications of calcium hydroxide in endodontics and dental traumatology. *Int. Endod. J.,* 44, 697-730.

Molander, A., Reit, C., Dahlen, G., and Kvist, T. (1998). Microbiological status of root-filled teeth with apical periodontitis. *Int. Endod. J.,* 31, 1-7.

Munoz-Price, L. S., Lolans, K., and Quinn, J. P. (2005). Emergence of resistance to daptomycin during treatment of vancomycin-resistant *Enterococcus faecalis* infection. *Clin. Infect. Dis.*, 41, 565-566.

Murray, B. E. (1992). Beta-lactamase-producing enterococci. *Antimicrob. Agents Chemother.*, 36, 2355-2359.

Naenni, N., Thoma, K., and Zehnder, M. (2004). Soft tissue dissolution capacity of currently used and potential endodontic irrigants. *J. Endod.*, 30, 785-787.

Nagayoshi, M., Kitamura, C., Fukuizumi, T., Nishihara, T., and Terashita, M. (2004). Antimicrobial effect of ozonated water on bacteria invading dentinal tubules. *J. Endod.*, 30, 778-781.

Nakajo, K., Komori, R., Ishikawa, S., Ueno, T., Suzuki, Y., Iwami, Y., and Takahashi, N. (2006). Resistance to acidic and alkaline environments in the endodontic pathogen *Enterococcus faecalis*. *Oral Microbiol. Immunol.*, 21, 283-288.

Nallapareddy, S. R., Qin, X., Weinstock, G. M., Hook, M., and Murray, B. E. (2000a). *Enterococcus faecalis* adhesin, ace, mediates attachment to extracellular matrix proteins collagen type IV and laminin as well as collagen type I. *Infect. Immun.*, 68, 5218-5224.

Nallapareddy, S. R., Singh, K. V., Duh, R. W., Weinstock, G. M., and Murray, B. E. (2000b). Diversity of ace, a gene encoding a microbial surface component recognizing adhesive matrix molecules, from different strains of *Enterococcus faecalis* and evidence for production of ace during human infections. *Infect. Immun.*, 68, 5210-5217.

Nallapareddy, S. R., Singh, K. V., Sillanpaa, J., Garsin, D. A., Hook, M., Erlandsen, S. L., and Murray, B. E. (2006). Endocarditis and biofilm-associated pili of *Enterococcus faecalis*. *J. Clin. Invest.*, 116, 2799-2807.

Narayanan, L. L., and Vaishnavi, C. (2010). Endodontic microbiology. *J. Conserv. Dent.*, 13, 233-239.

Neelakantan, P., Sanjeev, K., and Subbarao, C. V. (2007). Duration-dependent susceptibility of endodontic pathogens to calcium hydroxide and chlorhexidene gel used as intracanal medicament: an *in vitro* evaluation. *Oral Surg. Oral Med. Oral Pathol. Oral Radiol. Endod.*, 104, e138-141.

Nielsen, H. V., Flores-Mireles, A. L., Kau, A. L., Kline, K. A., Pinkner, J. S., Neiers, F., Normark, S., Henriques-Normark, B., Caparon, M. G., and Hultgren, S.J. (2013). Pilin and sortase residues critical for endocarditis- and biofilm-associated pilus biogenesis in *Enterococcus faecalis*. *J. Bacteriol.*, 195, 4484-4495.

Okino, L. A., Siqueira, E. L., Santos, M., Bombana, A. C., and Figueiredo, J. A. (2004). Dissolution of pulp tissue by aqueous solution of chlorhexidine digluconate and chlorhexidine digluconate gel. *Int. Endod. J.*, 37, 38-41.

Peciuliene, V., Maneliene, R., Balcikonyte, E., Drukteinis, S., and Rutkunas, V. (2008). Microorganisms in root canal infections: a review. *Stomatologija*, 10, 4-9.

Peciuliene, V., Reynaud, A. H., Balciuniene, I., and Haapasalo, M. (2001). Isolation of yeasts and enteric bacteria in root-filled teeth with chronic apical periodontitis. *Int. Endod. J.*, 34, 429-434.

Pinheiro, E. T., Gomes, B. P., Drucker, D. B., Zaia, A. A., Ferraz, C. C., and Souza-Filho, F. J. (2004). Antimicrobial susceptibility of *Enterococcus faecalis* isolated from canals of root filled teeth with periapical lesions. *Int. Endod. J.*, 37, 756-763.

Pinheiro, E. T., Gomes, B. P., Ferraz, C. C., Sousa, E. L., Teixeira, F. B., and Souza-Filho, F. J. (2003). Microorganisms from canals of root-filled teeth with periapical lesions. *Int. Endod. J.*, 36, 1-11.

Pinkston, K. L., Gao, P., Diaz-Garcia, D., Sillanpaa, J., Nallapareddy, S. R., Murray, B. E., and Harvey, B. R. (2011). The *Fsr* quorum-sensing system of *Enterococcus faecalis* modulates surface display of the collagen-binding MSCRAMM Ace through regulation of *gelE*. *J. Bacteriol.*, 193, 4317-4325.

Poptani, B., Sharaff, M., Archana, G., and Parekh, V. (2013). Detection of *Enterococcus faecalis* and Candida albicans in previously root-filled teeth in a population of Gujarat with polymerase chain reaction. *Contemp. Clin. Dent.*, 4, 62-66.

Portenier, I., Waltimo, T., Orstavik, D., and Haapasalo, M. (2005). The susceptibility of starved, stationary phase, and growing cells of *Enterococcus faecalis* to endodontic medicaments. *J. Endod.*, 31, 380-386.

Portenier, I., Waltimo, T., Orstavik, D., and Haapasalo, M. (2006). Killing of *Enterococcus faecalis* by MTAD and chlorhexidine digluconate with or without cetrimide in the presence or absence of dentine powder or BSA. *J. Endod.*, 32, 138-141.

Prajsnar, T. K., Renshaw, S. A., Ogryzko, N. V., Foster, S. J., Serror, P., and Mesnage, S. (2013). Zebrafish as a novel vertebrate model to dissect enterococcal pathogenesis. *Infect. Immun.*, 81, 4271-4279.

Qin, X., Singh, K. V., Weinstock, G. M., and Murray, B. E. (2000). Effects of *Enterococcus faecalis fsr* genes on production of gelatinase and a serine protease and virulence. *Infect. Immun.*, 68, 2579-2586.

Radcliffe, C. E., Potouridou, L., Qureshi, R., Habahbeh, N., Qualtrough, A., Worthington, H., and Drucker, D. B. (2004). Antimicrobial activity of varying concentrations of sodium hypochlorite on the endodontic microorganisms *Actinomyces israelii, A. naeslundii, Candida albicans* and *Enterococcus faecalis*. *Int. Endod. J.*, 37, 438-446.

Rakita, R. M., Vanek, N. N., Jacques-Palaz, K., Mee, M., Mariscalco, M. M., Dunny, G. M., Snuggs, M., Van Winkle, W. B., and Simon, S. I. (1999). *Enterococcus faecalis* bearing aggregation substance is resistant to killing by human neutrophils despite phagocytosis and neutrophil activation. *Infect. Immun.*, 67, 6067-6075.

Ramamurthy, N. S., Xu, J. W., Bird, J., Baxter, A., Bhogal, R., Wills, R., Watson, B., Owen, D., Wolff, M., and Greenwald, R. A. (2002). Inhibition of alveolar bone loss by matrix metalloproteinase inhibitors in experimental periodontal disease. *J. Periodontal. Res.*, 37, 1-7.

Rathnayake, I. U., Hargreaves, M., and Huygens, F. (2012). Antibiotic resistance and virulence traits in clinical and environmental *Enterococcus faecalis* and *Enterococcus faecium* isolates. *Syst. Appl. Microbiol.*, 35, 326-333.

Reynolds, P. E., Arias, C. A., and Courvalin, P. (1999). Gene vanXYC encodes D,D - dipeptidase (VanX) and D,D-carboxypeptidase (VanY) activities in vancomycin-resistant *Enterococcus gallinarum* BM4174. *Mol. Microbiol.*, 34, 341-349.

Ribeiro Sobrinho, A. P., Lanna, M. A., Farias, L. M., Carvalho, M. A., Nicoli, J. R., de Uzeda, M., and Vieira, L. Q. (2001). Implantation of bacteria from human pulpal necrosis and translocation from root canals in gnotobiotic mice. *J. Endod.*, 27, 605-609.

Rice, L. B., Bellais, S., Carias, L. L., Hutton-Thomas, R., Bonomo, R. A., Caspers, P., Page, M. G., and Gutmann, L. (2004). Impact of specific *pbp5* mutations on expression of beta-lactam resistance in *Enterococcus faecium*. *Antimicrob. Agents Chemother.*, 48, 3028-3032.

Rich, R. L., Kreikemeyer, B., Owens, R. T., LaBrenz, S., Narayana, S. V., Weinstock, G. M., Murray, B. E., and Hook, M. (1999). Ace is a collagen-binding MSCRAMM from *Enterococcus faecalis*. *J. Biol. Chem.*, 274, 26939-26945.

Rocas, I. N., Siqueira, J. F., Jr., and Santos, K. R. (2004). Association of *Enterococcus faecalis* with different forms of periradicular diseases. *J. Endod.*, 30, 315-320.

Rozdzinski, E., Marre, R., Susa, M., Wirth, R., and Muscholl-Silberhorn, A. (2001). Aggregation substance-mediated adherence of *Enterococcus faecalis* to immobilized extracellular matrix proteins. *Microb. Pathog.*, 30, 211-220.

Sava, I. G., Heikens, E., and Huebner, J. (2010). Pathogenesis and immunity in *enterococcal* infections. *Clin. Microbiol. Infect.*, 16, 533-540.

Sava, I. G., Zhang, F., Toma, I., Theilacker, C., Li, B., Baumert, T. F., Holst, O., Linhardt, R. J., and Huebner, J. (2009). Novel interactions of glycosaminoglycans and bacterial glycolipids mediate binding of enterococci to human cells. *J. Biol. Chem.*, 284, 18194-18201.

Sedgley, C. M., Lennan, S. L., and Clewell, D. B. (2004). Prevalence, phenotype and genotype of oral enterococci. *Oral Microbiol. Immunol.,* 19, 95-101.

Sedgley, C. M., Molander, A., Flannagan, S. E., Nagel, A. C., Appelbe, O. K., Clewell, D. B., and Dahlen, G. (2005). Virulence, phenotype and genotype characteristics of endodontic *Enterococcus* spp. *Oral Microbiol. Immunol.*, 20, 10-19.

Shankar, N., Lockatell, C. V., Baghdayan, A. S., Drachenberg, C., Gilmore, M. S., and Johnson, D. E. (2001). Role of *Enterococcus faecalis* surface protein Esp in the pathogenesis of ascending urinary tract infection. *Infect. Immun.*, 69, 4366-4372.

Shankar, V., Baghdayan, A. S., Huycke, M. M., Lindahl, G., and Gilmore, M. S. (1999). Infection-derived *Enterococcus faecalis* strains are enriched in *esp*, a gene encoding a novel surface protein. *Infect. Immun.*, 67, 193-200.

Shao, C., Shang, W., Yang, Z., Sun, Z., Li, Y., Guo, J., Wang, X., Zou, D., Wang, S., Lei, H., et al. (2012). LuxS-dependent AI-2 regulates versatile functions in *Enterococcus faecalis* V583. *J. Proteome. Res.*, 11, 4465-4475.

Sifaoui, F., Arthur, M., Rice, L., and Gutmann, L. (2001). Role of penicillin-binding protein 5 in expression of ampicillin resistance and peptidoglycan structure in *Enterococcus faecium*. *Antimicrob. Agents Chemother.*, 45, 2594-2597.

Sifri, C. D., Mylonakis, E., Singh, K. V., Qin, X., Garsin, D. A., Murray, B. E., Ausubel, F. M., and Calderwood, S. B. (2002). Virulence effect of *Enterococcus faecalis* protease genes and the quorum-sensing locus *fsr* in *Caenorhabditis elegans* and *mice*. *Infect. Immun.*, 70, 5647-5650.

Sillanpaa, J., Chang, C., Singh, K. V., Montealegre, M. C., Nallapareddy, S. R., Harvey, B. R., Ton-That, H., and Murray, B. E. (2013). Contribution of individual Ebp Pilus subunits of *Enterococcus faecalis* OG1RF to pilus biogenesis, biofilm formation and urinary tract infection. *PLoS ONE, 8*, e68813.

Sillanpaa, J., Nallapareddy, S. R., Houston, J., Ganesh, V. K., Bourgogne, A., Singh, K. V., Murray, B. E., and Hook, M. (2009). A family of fibrinogen-binding MSCRAMMs from *Enterococcus faecalis*. *Microbiology*, 155, 2390-2400.

Singh, K. V., Lewis, R. J., and Murray, B. E. (2009). Importance of the *epa* locus of *Enterococcus faecalis* OG1RF in a mouse model of ascending urinary tract infection. *J. Infect. Dis.*, 200, 417-420.

Singh, K. V., Nallapareddy, S. R., and Murray, B. E. (2007). Importance of the *ebp* (endocarditis- and biofilm-associated pilus) locus in the pathogenesis of *Enterococcus faecalis* ascending urinary tract infection. *J. Infect. Dis.*, 195, 1671-1677.

Singh, K. V., Nallapareddy, S. R., Nannini, E. C., and Murray, B. E. (2005). Fsr-independent production of protease(s) may explain the lack of attenuation of an *Enterococcus faecalis fsr* mutant versus a *gelE-sprE* mutant in induction of endocarditis. *Infect. Immun.*, 73, 4888-4894.

Singh, K. V., Nallapareddy, S. R., Sillanpaa, J., and Murray, B. E. (2010). Importance of the collagen adhesin ace in pathogenesis and protection against *Enterococcus faecalis* experimental endocarditis. *PLoS Pathog.*, 6, e1000716.

Singh, K. V., Weinstock, G. M., and Murray, B. E. (2002). An *Enterococcus faecalis* ABC homologue (Lsa) is required for the resistance of this species to clindamycin and quinupristin-dalfopristin. *Antimicrob. Agents Chemother.*, 46, 1845-1850.

Siqueira, J. F., Jr., Batista, M. M., Fraga, R. C., and de Uzeda, M. (1998). Antibacterial effects of endodontic irrigants on black-pigmented gram-negative anaerobes and facultative bacteria. *J. Endod.*, 24, 414-416.

Siqueira, J. F., Jr., Guimaraes-Pinto, T., and Rocas, I. N. (2007). Effects of chemomechanical preparation with 2.5% sodium hypochlorite and intracanal medication with calcium hydroxide on cultivable bacteria in infected root canals. *J. Endod.*, 33, 800-805.

Siqueira, J. F., Jr., and Lopes, H. P. (1999). Mechanisms of antimicrobial activity of calcium hydroxide: a critical review. *Int. Endod. J.*, 32, 361-369.

Siqueira, J. F., Jr., Machado, A. G., Silveira, R. M., Lopes, H. P., and de Uzeda, M. (1997). Evaluation of the effectiveness of sodium hypochlorite used with three irrigation methods in the elimination of *Enterococcus faecalis* from the root canal, in vitro. *Int. Endod. J.*, 30, 279-282.

Siqueira, J. F., Jr., and Rocas, I. N. (2004). Polymerase chain reaction-based analysis of microorganisms associated with failed endodontic treatment. *Oral Surg. Oral Med. Oral Pathol. Oral Radiol. Endod.*, 97, 85-94.

Siqueira, J. F., Jr., Rocas, I. N., Favieri, A., and Lima, K. C. (2000). Chemomechanical reduction of the bacterial population in the root canal after instrumentation and irrigation with 1%, 2.5%, and 5.25% sodium hypochlorite. *J. Endod.*, 26, 331-334.

Sjogren, U., Hagglund, B., Sundqvist, G., and Wing, K. (1990). Factors affecting the long-term results of endodontic treatment. *J. Endod.*, 16, 498-504.

Sobrinho, A. P., Barros, M. H., Nicoli, J. R., Carvalho, M. A., Farias, L. M., Bambirra, E. A., Bahia, M. G., and Vieira, E. C. (1998). Experimental root canal infections in conventional and germ-free mice. *J. Endod.*, 24, 405-408.

Soell, M., Elkaim, R., and Tenenbaum, H. (2002). Cathepsin C, matrix metalloproteinases, and their tissue inhibitors in gingiva and gingival crevicular fluid from periodontitis-affected patients. *J. Dent. Res.*, 81, 174-178.

Souto, R., and Colombo, A. P. (2008). Prevalence of *Enterococcus faecalis* in subgingival biofilm and saliva of subjects with chronic periodontal infection. *Arch. Oral Biol.*, 53, 155-160.

Strickertsson, J. A., Desler, C., Martin-Bertelsen, T., Machado, A. M., Wadstrom, T., Winther, O., Rasmussen, L. J., and Friis-Hansen, L. (2013). *Enterococcus faecalis* infection causes inflammation, intracellular oxphos-independent ROS production, and DNA damage in human gastric cancer cells. *PLoS ONE*, 8, e63147.

Stuart, C. H., Schwartz, S. A., Beeson, T. J., and Owatz, C. B. (2006). *Enterococcus faecalis*: its role in root canal treatment failure and current concepts in retreatment. *J. Endod.*, 32, 93-98.

Sunde, P. T., Olsen, I., Debelian, G. J., and Tronstad, L. (2002). Microbiota of periapical lesions refractory to endodontic therapy. *J. Endod.*, 28, 304-310.

Sundqvist, G., Figdor, D., Persson, S., and Sjogren, U. (1998). Microbiologic analysis of teeth with failed endodontic treatment and the outcome of conservative re-treatment. *Oral Surg. Oral Med. Oral Pathol. Oral Radiol. Endod.*, 85, 86-93.

Sussmuth, S. D., Muscholl-Silberhorn, A., Wirth, R., Susa, M., Marre, R., and Rozdzinski, E. (2000). Aggregation substance promotes adherence, phagocytosis, and intracellular survival of *Enterococcus faecalis* within human macrophages and suppresses respiratory burst. *Infect. Immun.*, 68, 4900-4906.

Teixeira, N., Varahan, S., Gorman, M. J., Palmer, K. L., Zaidman-Remy, A., Yokohata, R., Nakayama, J., Hancock, L. E., Jacinto, A., Gilmore, M. S., *et al.* (2013). *Drosophila* host model reveals new *enterococcus faecalis* quorum-sensing associated virulence factors. *PLoS ONE*, 8, e64740.

Tendolkar, P. M., Baghdayan, A. S., Gilmore, M. S., and Shankar, N. (2004). *Enterococcal* surface protein, Esp, enhances biofilm formation by *Enterococcus faecalis. Infect. Immun.*, 72, 6032-6039.

Tendolkar, P. M., Baghdayan, A. S., and Shankar, N. (2006). Putative surface proteins encoded within a novel transferable locus confer a high-biofilm phenotype to *Enterococcus faecalis. J. Bacteriol.*, 188, 2063-2072.

Teng, F., Jacques-Palaz, K. D., Weinstock, G. M., and Murray, B. E. (2002). Evidence that the enterococcal polysaccharide antigen gene (*epa*) cluster is widespread in *Enterococcus faecalis* and influences resistance to phagocytic killing of *E. faecalis. Infect. Immun.*, 70, 2010-2015.

Teng, F., Singh, K. V., Bourgogne, A., Zeng, J., and Murray, B. E. (2009). Further characterization of the *epa* gene cluster and Epa polysaccharides of *Enterococcus faecalis. Infect. Immun.*, 77, 3759-3767.

Theilacker, C., Sanchez-Carballo, P., Toma, I., Fabretti, F., Sava, I., Kropec, A., Holst, O., and Huebner, J. (2009). Glycolipids are involved in biofilm accumulation and prolonged bacteraemia in *Enterococcus faecalis. Mol. Microbiol.*, 71, 1055-1069.

Toledo-Arana, A., Valle, J., Solano, C., Arrizubieta, M. J., Cucarella, C., Lamata, M., Amorena, B., Leiva, J., Penades, J. R., and Lasa, I. (2001). The enterococcal surface protein, Esp, is involved in *Enterococcus faecalis* biofilm formation. *Appl. Environ. Microbiol.*, 67, 4538-4545.

Uttley, A. H., Collins, C. H., Naidoo, J., and George, R. C. (1988). Vancomycin-resistant enterococci. *Lancet*, 1, 57-58.

Vahdaty, A., Pitt Ford, T. R., and Wilson, R. F. (1993). Efficacy of chlorhexidine in disinfecting dentinal tubules *in vitro. Endod. Dent. Traumatol.*, 9, 243-248.

Valadares, M. A., Soares, J. A., Nogueira, C. C., Cortes, M. I., Leite, M. E., Nunes, E., and Silveira, F. F. (2011). The efficacy of a cervical barrier in preventing microleakage of *Enterococcus faecalis* in endodontically treated teeth. *Gen. Dent.*, 59, e32-37.

Vanek, N. N., Simon, S. I., Jacques-Palaz, K., Mariscalco, M. M., Dunny, G.M., and Rakita, R.M. (1999). *Enterococcus faecalis* aggregation substance promotes opsonin-independent binding to human neutrophils via a complement receptor type 3-mediated mechanism. *FEMS Immunol. Med. Microbiol.*, 26, 49-60.

Vianna, M. E., Gomes, B. P., Berber, V. B., Zaia, A. A., Ferraz, C. C., and de Souza-Filho, F. J. (2004). In vitro evaluation of the antimicrobial activity of chlorhexidine and sodium hypochlorite. *Oral Surg. Oral Med. Oral Pathol. Oral Radiol. Endod.*, 97, 79-84.

Wang, Q. Q., Zhang, C. F., Chu, C. H., and Zhu, X. F. (2012). Prevalence of *Enterococcus faecalis* in saliva and filled root canals of teeth associated with apical periodontitis. *Int. J. Oral Sci.*, 4, 19-23.

Wang, X., Allen, T. D., May, R. J., Lightfoot, S., Houchen, C. W., and Huycke, M.M. (2008). *Enterococcus faecalis* induces aneuploidy and tetraploidy in colonic epithelial cells through a bystander effect. *Cancer Res.*, 68, 9909-9917.

Wang, X., and Huycke, M. M. (2007). Extracellular superoxide production by *Enterococcus faecalis* promotes chromosomal instability in mammalian cells. *Gastroenterology*, 132, 551-561.

Werner, G., Strommenger, B., and Witte, W. (2008). Acquired vancomycin resistance in clinically relevant pathogens. *Future Microbiol.*, 3, 547-562.

Williams, N. B., Forbes, M. A., Blau, E., and Eickenberg, C. F. (1950). A study of the simultaneous occurrence of Enterococci, Lactobacilli, and yeasts in saliva from human beings. *J. Dent. Res.*, 29, 563-570.

Wilson, D. N., Schluenzen, F., Harms, J. M., Starosta, A. L., Connell, S. R., and Fucini, P. (2008). The oxazolidinone antibiotics perturb the ribosomal peptidyl-transferase center and effect tRNA positioning. *Proc. Natl. Acad. Sci. U S A,* 105, 13339-13344.

Yesilsoy, C., Whitaker, E., Cleveland, D., Phillips, E., and Trope, M. (1995). Antimicrobial and toxic effects of established and potential root canal irrigants. *J. Endod.*, 21, 513-515.

Zehnder, M. (2006). Root canal irrigants. *J. Endod.*, 32, 389-398.

Zehnder, M., and Guggenheim, B. (2009a). The mysterious appearance of enterococci in filled root canals. *Int. Endod. J.*, 42, 277-287.

Zehnder, M., and Guggenheim, B. (2009b). The mysterious appearance of enterococci in filled root canals. *Int. Endod. J.*, 42, 277-287.

Zhu, X., Wang, Q., Zhang, C., Cheung, G. S., and Shen, Y. (2010). Prevalence, phenotype, and genotype of *Enterococcus faecalis* isolated from saliva and root canals in patients with persistent apical periodontitis. *J. Endod.*, 36, 1950-1955.

Zoletti, G. O., Siqueira, J. F., Jr., and Santos, K. R. (2006). Identification of *Enterococcus faecalis* in root-filled teeth with or without periradicular lesions by culture-dependent and-independent approaches. *J. Endod.*, 32, 722-726.

INDEX

A

Abraham, 84
access, 152
acetylation, 36
acid, 4, 16, 22, 23, 26, 27, 33, 37, 74, 75, 84, 86, 128, 132, 152, 154, 155
acidic, 159
acquired characteristics, 35
active site, 35, 141
active transport, 143
adaptability, 83
adaptation, 82
adenine, 37
adenosine, 39
adhesion, 3, 4, 7, 12, 13, 32, 33, 40, 70, 72, 94, 118, 128, 138, 139, 140, 156, 157
adhesion properties, 72
adjustment, 38
adsorption, 118
adverse conditions, xi, 10, 73, 123, 124
agar, 24, 25, 46, 96, 111, 112, 150, 151
age, 31, 58
aggregation, vii, 1, 4, 5, 6, 7, 11, 12, 15, 32, 33, 50, 72, 73, 78, 79, 86, 89, 105, 139, 154, 160, 164
agriculture, 97
alanine, 42, 142, 143
albumin, 71
algae, 26
alkalinity, 119
allele, 39, 47
allergy, 9, 142
alters, 70, 110, 118
AME, 36
amine(s), xi, 123, 127, 128, 131, 133
amino, xi, 3, 4, 5, 32, 37, 86, 124, 128
amino acid(s), xi, 3, 4, 5, 32, 37, 124, 128

aminoglycoside(s), viii, x, 9, 20, 30, 34, 35, 36, 53, 55, 58, 59, 61, 62, 64, 66, 93, 95, 96, 97, 98, 100, 101, 104, 105, 106, 107, 141, 142, 143, 144
anaerobe, 138
anaerobic bacteria, 42, 121, 145
anatomy, 148
aneuploidy, 164
animal husbandry, 41, 81
antagonism, 26
antibiotic resistance, viii, xi, 5, 6, 8, 9, 10, 12, 13, 15, 19, 20, 26, 47, 48, 53, 54, 56, 57, 58, 62, 65, 67, 72, 73, 78, 79, 80, 81, 85, 89, 90, 92, 105, 107, 123, 127, 128, 129, 131, 132, 133, 134, 135, 141, 147, 153, 156
antibody, 140
antigen, 5, 14, 33, 86, 94, 128, 134, 140, 163
anti-inflammatory drugs, 28
antimicrobial therapy, 20, 100
antimicrobials, viii, ix, x, 9, 21, 22, 29, 30, 33, 34, 35, 38, 41, 42, 45, 79, 85, 94, 95, 97, 98, 99, 100, 102, 103, 132, 135, 149
apex, 148
apoptosis, 71, 84
appointments, 110, 119, 147, 149
Arabidopsis thaliana, 145, 157
Argentina, 50, 60, 93, 97, 98, 104, 109, 120
aromatic rings, 21
arrest, 158
arthritis, 60
aseptic, 149
Asia, 39, 41
assessment, 57, 85, 134
asymptomatic, ix, 70, 146
atmosphere, 57
ATP, 7, 37, 143
attachment, 84, 157, 159
autolysis, 6, 8, 16

B

Bacillus subtilis, 22
bacteraemia, vii, 1, 2, 15, 54, 63, 86, 105, 163
bacteremia, viii, 19, 29, 31, 37, 48, 53, 54, 100, 101, 102, 103, 106, 107, 127, 138, 140, 143, 157
bacterial cells, 7, 110, 145
bacterial infection, viii, 19, 65, 83, 100, 148
bacterial pathogens, 48
bacterial strains, x, 94, 103, 124, 127
bacteriocins, 125, 129, 133, 135
bacteriolysis, 71
bacteriostatic, 39, 95, 118, 143, 151, 152
bacterium, vii, ix, x, xi, 66, 69, 70, 81, 109, 110, 137, 147, 152
barriers, 30, 71
base, 99, 119, 152
base pair, 99
BD, 85, 87, 105, 106
beef, 92
BI, 90
bile, 34, 46, 94
bioavailability, 38
biocompatibility, 110, 153, 156
biodiversity, 92
biofilm production, viii, 6, 7, 8, 11, 29, 84
biological activities, 21, 26
biomass, 78
biopolymers, 33
biopreservation, 107, 124, 129, 130, 133
biosynthesis, 8, 15, 17, 40, 140, 144
biotic, 7
birds, 94
blood, 7, 13, 17, 49, 85, 87, 98, 102, 106, 112, 140, 157
blood cultures, 98, 102, 106
blood stream, 140
bloodstream, 48, 52, 59
body fluid, 98
bone, 41, 48, 55, 87, 139, 160
bone marrow, 41, 48, 55
bone marrow transplant, 41, 48, 55
bone resorption, 87, 139
brain, 6
Brazil, 48, 49, 50, 51, 54, 59, 60, 61, 64, 65, 66, 67, 88, 101, 102, 104
breakdown, 150
breeding, 97
broad-spectrum antimicrobial agents, viii, 29
burn, 53
bystander effect, 164

C

Ca^{2+}, 150
calcification, 150
calcium, x, xi, 39, 109, 112, 113, 119, 120, 121, 143, 147, 149, 150, 151, 153, 155, 157, 158, 159, 162
campaigns, 45
canals, 7, 110, 113, 118, 119, 120, 121, 138, 139, 144, 146, 147, 148, 149, 150, 152, 153, 154, 155, 159, 160, 162, 164
cancer, 27, 28, 139, 163
cancer cells, 163
candida, 129
candidates, 28, 126, 131
capsule, 5, 71, 89
carbohydrate(s), 73, 74, 140
carbohydrate metabolism, 73
carbon, 27, 118
cartilage, 6
cascades, 73
casein, 6, 125, 126, 139
cataract, 127
catheter, 30, 31, 41
cation, 151
cattle, 39, 97
CCR, 120
CDC, 45, 49, 101
cell cycle, 158
cell death, 39, 95, 151
cell division, 143
cell membranes, 94
cell surface, 4, 71, 73, 75, 85, 140
cellulitis, 147, 148
challenges, 12, 17, 34, 47, 138, 149
cheese, 15, 79, 92, 97, 98, 107, 124, 125, 126, 128, 129, 130, 132, 134, 135
chemical(s), viii, 19, 20, 21, 26, 52, 71, 130, 144, 147, 154
chemokines, 71
chemotherapy, 17, 26, 27, 28
Chicago, 102
chicken, 44, 60, 90, 92, 97, 98, 103
Chile, 98
China, 39, 44
chirality, 20
chloramphenicol, viii, 9, 30, 34, 38, 79, 81
chloramphenicol resistance, 79
chlorine, 99
cholecystectomy, 52
chromosomal instability, 164
chromosome, 6, 7, 8, 34, 38, 43, 46, 56
City, 65, 98
classes, viii, 9, 26, 30, 41, 81

classification, 133
cleaning, x, 45, 109, 148
clinical syndrome, viii, 29
clone, 48, 50, 91
cloning, 87
clusters, 20, 33, 43, 44, 49, 70, 82, 91, 104, 140
coccus, vii, ix, 69, 117, 133, 134
coding, 9, 98
collagen, 4, 6, 15, 33, 62, 74, 86, 87, 88, 139, 140, 144, 145, 147, 149, 156, 159, 160, 161, 162
Colombia, 104
colon, 4, 99
colonization, 2, 3, 12, 33, 34, 41, 42, 45, 48, 51, 54, 62, 66, 68, 94, 99, 116
commensal, vii, viii, ix, xi, 4, 6, 10, 19, 34, 47, 69, 70, 71, 72, 73, 78, 81, 82, 83, 84, 85, 89, 137, 138, 139, 158
commensalism, 155
communication, xii, 52, 73, 85, 137, 139, 144, 154, 157
communication systems, 144, 157
community(s), vii, 1, 2, 11, 31, 41, 45, 50, 53, 59, 73, 78, 83, 88, 91, 100, 101, 154
comparative analysis, 89, 94
competitors, 139
compilation, 94
complement, ix, 33, 69, 71, 74, 75, 84, 164
complexity, 13
complications, 31, 127
composition, 70
compounds, viii, 6, 7, 19, 20, 21, 22, 23, 24, 25, 26, 39, 125, 126, 134, 142, 143
conjugation, 4, 6, 13, 21, 32, 34, 43, 52, 79, 98, 139, 141, 154
connective tissue, 7, 113
Consensus, 64
consumption, 63, 90
contamination, xi, 13, 79, 123, 124, 126, 138, 149
controversial, xi, 42, 123, 139, 151
correlation, xi, 43, 123, 127, 129, 131, 139
correlation analysis, 131
cost, 41, 46
cotton, 114
Croatia, 125, 133
crown(s), 113, 148
CSF, 71
Cuba, 61
cues, 70, 72
cultural differences, 125
culture, xi, 15, 17, 22, 46, 51, 73, 84, 85, 102, 109, 110, 112, 115, 131, 134, 144, 164
culture conditions, 73
culture medium, xi, 109, 112, 115

cure, 31, 143
CV, 87
cystitis, 60
cytokines, 70, 71, 84, 87
cytolysin, vii, ix, x, 1, 7, 10, 12, 50, 55, 69, 79, 87, 90, 93, 94, 98, 99, 102, 105, 128, 156
cytoplasm, 118, 119
cytotoxicity, 113
Czech Republic, 57

D

dairy industry, 125
damages, 13, 157
database, 47, 48, 82, 97
decay, 130
defence, 127
deficiency, 8, 149
degradation, 7, 32, 124, 125, 126
denaturation, 150
dendrogram, 82
Denmark, 91, 97, 104, 106
dental infections, vii, 2
dentin, 4, 112, 113, 118, 119, 120, 121, 140, 149, 153, 156, 157
deoxyribonucleic acid, 11, 90
deprivation, xii, 137
depth, 110, 151
destruction, 39, 143
detection, ix, 42, 46, 51, 60, 61, 64, 66, 68, 69, 72, 85, 94, 107, 128, 131, 132, 146, 155
developing countries, 46
diabetes, 41, 60
diarrhea, 66
diffusion, 24, 25, 96, 110, 111, 112, 150, 151
digestion, 46
Discitis, 60
diseases, vii, x, 1, 12, 17, 20, 31, 32, 58, 93, 94, 102, 103, 125, 138, 161
disinfection, 99, 121, 148
dispersion, 57
dissociation, 7, 110, 150
distilled water, xi, 109, 111, 113, 114, 116, 117, 118, 119
distribution, 15, 26, 28, 104, 129, 134
divergence, 44
diversity, xi, 4, 20, 48, 51, 61, 64, 89, 91, 101, 104, 123, 124, 129, 131, 133
DNA, 6, 13, 16, 28, 32, 38, 46, 47, 49, 56, 64, 65, 73, 78, 83, 88, 89, 94, 139, 150, 157, 163
DNA damage, 163
dominance, 125, 129, 138
donors, 151

draft, 89
dressing material, 150
dressings, xi, 109, 112, 120
drinking water, 13
Drosophila, 163
drug discovery, viii, 19, 21, 26
drug resistance, 14, 20, 90
drugs, viii, ix, 19, 22, 26, 27, 28, 30, 38, 40, 42, 47, 95, 100, 103, 138, 142, 154, 156
D-serine, 62, 143
duodenum, 4

E

ecology, 11, 13, 104, 105
economic activity, 97
editors, 120
effluent, 99
electron, xi, 109
electron microscopy, xi, 109
electrophoresis, 46, 51, 54, 58, 60, 65, 66, 89, 97, 107
elongation, 40, 71
elucidation, 21
e-mail, 1
emergency, 61, 99
EMS, 107
encoding, xi, 3, 5, 6, 7, 8, 10, 16, 32, 37, 42, 58, 62, 70, 77, 78, 81, 86, 87, 89, 90, 95, 123, 139, 140, 159, 161
endocarditis, vii, viii, x, xi, 1, 2, 4, 5, 6, 8, 10, 11, 14, 17, 19, 20, 29, 31, 32, 33, 48, 50, 53, 55, 57, 60, 89, 93, 97, 100, 101, 105, 128, 134, 137, 138, 139, 140, 141, 142, 143, 144, 145, 153, 154, 155, 156, 159, 162
end-stage renal disease, 54
energy, 145
England, 41, 56
enlargement, 153
enterococcal infections, vii, viii, 1, 2, 9, 15, 29, 30, 34, 35, 37, 39, 40, 41, 45, 60, 61, 63, 81, 88, 94, 100, 102, 103, 141, 142, 161
environment(s), vii, viii, ix, 1, 2, 8, 10, 12, 29, 30, 31, 34, 41, 45, 47, 70, 71, 79, 82, 83, 97, 99, 110, 114, 119, 124, 127, 128, 139, 147, 149, 157, 159
environmental conditions, 2, 71, 94
environmental contamination, 45
enzyme(s), vii, ix, x, 1, 6, 7, 32, 35, 36, 38, 42, 65, 69, 70, 93, 95, 96, 104, 106, 107, 110, 118, 128, 142, 143
EPA, 140
epidemic, 45, 89, 101
epidemiologic, 64, 67

epidemiological investigations, 51
epidemiology, 11, 45, 47, 48, 49, 50, 52, 55, 58, 59, 60, 61, 62, 65, 66, 67, 82, 97, 101, 104, 107, 134, 156
epinephrine, 145
epithelia, 32
epithelial cells, vii, 1, 16, 32, 74, 86, 140, 164
epithelium, 32, 139
equilibrium, 111
equipment, 46, 79
erythrocytes, 7, 75, 139
espionage, 157
ethanol, 125
etiology, 55, 58, 63
eukaryotic, 33, 128, 157
eukaryotic cell, 33, 128
Europe, 30, 35, 37, 39, 41, 52, 60, 91, 96, 106
evidence, x, 11, 31, 52, 57, 81, 83, 97, 100, 109, 118, 147, 154, 159
evolution, 13, 22, 28, 47, 78, 83, 91
exclusion, ix, 70
exercise, 112
exertion, 99
experimental condition, 131
exposure, 71, 141, 143, 147
expulsion, 38
extracellular matrix, xii, 33, 73, 75, 86, 137, 139, 140, 145, 159, 161
extraction, 46, 94
extracts, 26

F

facilitators, 91
families, 95
farms, 97
fatty acids, 71, 84
faulty restorations, 147
FDA, 39, 40
feces, 2, 11, 50, 82, 94, 97, 98, 99, 104
fermentation, 24, 94, 124, 125, 134
fibers, 149
fibrin, 4, 12
fibrinogen, 74, 140, 161
fidelity, 95
filling materials, 156
fitness, 72
flavour, 124, 125, 126, 132
flora, 107, 117, 132, 138, 147
fluid, 139, 147, 148, 162
fluoroquinolones, 38
folic acid, 143

food, vii, viii, ix, xi, 1, 2, 10, 11, 12, 13, 15, 19, 41, 52, 58, 61, 79, 81, 90, 92, 93, 97, 98, 99, 101, 104, 105, 107, 123, 124, 125, 126, 127, 128, 129, 130, 131, 132, 133, 134, 135, 138, 147, 157
food chain, 81, 97, 98, 127
food industry, xi, 124, 127, 128
food poisoning, 127
food production, 127, 131, 135, 138
food products, vii, 1, 2
food safety, 12, 133
food spoilage, xi, 123, 126, 129, 130
Ford, 163
formation, 4, 7, 8, 10, 11, 14, 16, 17, 32, 33, 35, 42, 44, 72, 73, 74, 75, 84, 85, 86, 87, 88, 112, 120, 124, 126, 139, 140, 141, 145, 150, 154, 155, 157, 158, 161, 163
fragments, 34, 46, 47
France, 41, 95, 97
fruits, viii, 19
functional food, xi, 123
fungi, 22, 26, 149
fungus, 23, 24, 27
fusion, 36

G

gastrointestinal tract, xi, 2, 30, 45, 81, 123, 144
gel, x, xi, 51, 54, 58, 60, 65, 66, 72, 76, 77, 89, 97, 107, 109, 110, 111, 112, 113, 114, 116, 118, 119, 120, 144, 155, 159
gene amplification, 99
gene expression, 8, 53, 73, 83, 85
gene pool, 81, 83
gene transfer, 70, 73, 78, 83, 89, 100, 129, 139
genetic diversity, 12, 107, 132
genetic information, 34
genetics, 132
genitourinary tract, 30, 38
genome, ix, 20, 47, 69, 70, 72, 78, 79, 83, 87, 89, 91, 139, 140, 145
genomic regions, 83
genomics, 88, 89
genotype, ix, 30, 44, 49, 64, 83, 161, 164
genotyping, 82
genus, viii, ix, 29, 31, 32, 34, 35, 38, 63, 93, 94, 104, 124, 125, 134
Germany, 15, 67, 134
gingival, 139, 147, 162
gland, 57
glucose, 78, 88, 94, 126
glycans, 84
glycerin, 113
glycol, x, 109, 111, 112, 152

glycopeptides, viii, 9, 13, 30, 34, 35, 40, 42, 43, 53, 66, 95, 98, 100, 103, 106, 142
glycosaminoglycans, 161
GRAS, 126
grazing, 124
Greece, 41, 51, 102
grouping, 133
growth, xi, 2, 24, 25, 30, 32, 35, 37, 41, 44, 70, 71, 73, 85, 87, 94, 96, 97, 109, 110, 111, 112, 115, 118, 130, 131, 133, 141, 150, 151, 155
growth factor, 71
growth rate, 73, 94
growth temperature, 94
guidelines, 45, 48, 53, 64

H

habitat(s), 2, 81
harbors, 139, 145
health, x, xi, 6, 20, 31, 45, 94, 101, 123, 124, 133, 138, 153
health care, 138
heart failure, 31
hematology, 45
hemocoel, 158
hemodialysis, 41, 54
hepatic failure, 38
heterogeneity, 41, 61, 73, 83, 85
history, 31, 65, 138
HM, 106
homeostasis, 5
horses, 32
hospitalization, ix, x, 30, 39, 41, 94, 103
hospitalized patients, viii, 29, 47, 50, 52, 53, 54, 82, 96, 101
host, vii, ix, xii, 1, 3, 4, 5, 6, 10, 13, 20, 32, 33, 35, 52, 57, 69, 70, 71, 74, 75, 79, 86, 94, 100, 101, 110, 128, 137, 139, 140, 141, 145, 147, 149, 154, 157, 163
hotspots, 21
human health, vii, ix, 69
human neutrophils, 15, 160, 164
hybridization, 89
hydrogen, 13, 130, 139, 143, 157
hydrogen bonds, 143
hydrogen peroxide, 13, 130, 139, 157
hydrolysis, 6, 42, 43, 143
hydrophobicity, 4, 12, 73
hydroxide, x, xi, 109, 110, 112, 113, 119, 120, 121, 147, 149, 150, 151, 153, 155, 158, 159, 162
hydroxyapatite, 113, 151
hydroxyl, 36, 38
hygiene, 45

hypersensitivity, 143

I

identification, 11, 32, 43, 46, 49, 52, 56, 58, 63, 72, 85, 87, 89, 90, 91, 104, 132, 146
identity, 32, 43
IL-8, 71
immune activation, 70
immune response, vii, ix, 1, 32, 70, 71, 94, 103, 147
immune system, 5
immunity, 7, 16, 63, 161
immunization, 140
immunocompromised, 12, 30, 128
immunomodulatory, xii, 103, 137
immunostimulatory, 70, 84
immunosuppression, 2
in vitro, x, 16, 33, 39, 40, 44, 87, 97, 109, 110, 115, 117, 119, 120, 140, 143, 150, 151, 155, 156, 158, 159, 162, 163
in vivo, ix, 10, 16, 27, 32, 69, 99, 139, 140, 143
incidence, 17, 31, 41, 58, 63, 80, 132, 134, 157
incompatibility, 56
incubation period, 118
India, 61
individuals, 2, 6, 32, 41, 45, 100
inducer, 43, 145
induction, 44, 73, 89, 139, 156, 162
industries, 21
infants, 12
inflammation, 3, 6, 70, 84, 147, 148, 163
ingredients, 135
inhibition, ix, x, 24, 25, 70, 87, 96, 109, 110, 111, 112, 113, 118, 141, 151
inhibitor, 141
initiation, 38
injury(s), 6, 7, 30
inoculation, 126
inoculum, 42, 112, 141, 142
insects, 94, 145
insertion, 36, 39, 51, 78
integration, 46, 83
integrity, 110, 118
intensive care unit, 12, 50, 54, 57, 62, 65, 67, 68, 100, 102
internalization, 16, 32, 74, 86
intestinal microbiota, vii, 1, 2, 94, 99, 100, 105
intestinal tract, vii, xi, 13, 59, 99, 100, 137, 138
intestine, 58, 99
invertebrates, 26
ions, 119, 150
Iran, 85
Ireland, 41

irrigation, 110, 121, 149, 150, 152, 162
islands, 32, 70, 78, 83, 89
isolation, ix, 26, 30, 41, 46, 50, 51, 52, 59, 67, 90, 132, 147
isoprene, 22
Israel, 137
Italy, 49, 67, 90, 107

J

Japan, 44, 58, 60, 101

K

kidney, 54
kill, 153, 158
kinetics, 133
Korea, 44, 60, 97, 103

L

lactic acid, vii, ix, 69, 94, 124, 133, 134, 135
Latin America, 41, 48, 62
lead, 26, 34, 38, 39, 43, 71, 110, 148, 152
leakage, 39, 118, 147, 149
lecithin, 75
lesions, 110, 146, 148, 152, 159, 160, 163, 164
leucine, 33, 74, 86
leukocytes, 4, 70, 140
lichen, 26, 27
ligament, 113, 147, 149, 154
ligand, 145
light, 96
lipids, 6
lipoproteins, 11, 70
liver, 17, 55
liver cirrhosis, 55
liver transplant, 17
livestock, 97
loci, 5, 33, 82
locus, 6, 8, 14, 16, 33, 66, 88, 97, 156, 161, 162, 163
LTA, 33, 70, 71, 74
lymph, 138, 140
lymph node, 138, 140
lymphocytes, 149, 158
lysis, 7, 73, 88, 128, 140
lysozyme, 71

Index

M

macrophages, 4, 7, 15, 33, 64, 73, 75, 87, 88, 139, 163
Mainland China, 68
majority, vii, viii, 1, 2, 29, 37, 47, 72, 142
malaria, 45
Malaysia, 101, 107
mammalian cells, 164
mammalian tissues, 127
man, 95, 98, 99
management, ix, 30, 31, 45, 83
manganese, 5
mastitis, 11
materials, vii, 1, 2, 8, 120, 148, 151, 152, 153, 156, 158
matrix, 7, 33, 75, 128, 140, 159, 160, 162
matrix metalloproteinase, 160, 162
matter, 150
MB, 90
meat, 10, 12, 14, 15, 44, 49, 57, 60, 63, 79, 90, 91, 98, 99, 103, 124, 130
media, 46, 112, 118
medical, viii, 8, 11, 13, 16, 20, 30, 52, 62, 63, 71, 89, 133, 134
medical science, 71
medication, 113, 114, 118, 119, 120, 162
medicine, x, 9, 93, 127, 129, 131, 138
Mediterranean, 125
Mediterranean countries, 125
membranes, 157
meningitis, viii, x, 19, 56, 60, 67, 93, 138, 143
messengers, 70
meta-analysis, 52
metabolism, 26, 105, 110, 145
metabolites, 27
metabolizing, 149
metabolome, 83
methodology, 46, 119
methyl group, 42
methylation, 37
mice, 34, 58, 87, 99, 105, 140, 145, 157, 158, 160, 161, 162
microbial cells, 150
microbial community, 130
microbiota, vii, viii, ix, 1, 2, 29, 30, 42, 45, 46, 70, 93, 94, 99, 100, 102, 105, 107, 145
microorganism(s), viii, xii, 7, 10, 26, 27, 29, 42, 65, 112, 113, 117, 118, 119, 124, 126, 128, 130, 134, 137, 139, 146, 147, 149, 160, 162
Ministry of Education, 132
mitochondria, 158
MMP, 158
MMP-2, 158
MMP-9, 158
models, x, 6, 32, 33, 34, 72, 85, 93, 95, 99, 138, 139, 140, 145, 150
molecular biology, 48, 53, 54, 56, 58, 67, 135
molecular mass, 3, 126
molecular weight, 3, 6, 99, 118
molecules, viii, 8, 9, 11, 19, 20, 21, 23, 25, 26, 29, 32, 33, 34, 37, 40, 42, 70, 73, 75, 90, 139, 140, 142, 159
monolayer, 33
morbidity, 2, 99, 100
morphology, 156
mortality, 2, 20, 31, 35, 41, 46, 52, 55, 100, 101, 102
mortality rate, 20, 31, 101
motif, 4, 33
MR, 88, 120
mucin, 139
mucous membrane(s), 111
multi-drug resistant bacteria, viii, 19
multiple factors, 78
mutagen, 87
mutagenesis, 8
mutant, 8, 63, 162
mutation(s), viii, 19, 35, 38, 39, 64, 78, 141, 142, 143, 160

N

NaCl, 94, 133
NADH, 34, 87
nanoparticles, 152
National Academy of Sciences, 10, 11, 12
natural compound, 27, 39
natural food, 124
natural isolates, xi, 123, 131
necrosis, 71, 150, 160
neonatal sepsis, 2
nervous system, 31
Netherlands, 28, 53, 66
neutral, 126
neutropenia, 102
neutrophils, 7, 15, 32, 74
New England, 14, 27
New Zealand, 133
nitric oxide, 84
nitrogen, 71
North America, 39, 52, 156
nosocomial infections, vii, viii, ix, xi, 1, 2, 3, 10, 19, 30, 41, 50, 63, 69, 70, 90, 100, 102, 127, 137, 138
nucleic acid, 6, 7, 46
nucleotide sequence, 89, 90
nucleotides, 3, 143

nutrient(s), 6, 7, 71, 110, 112, 118, 149
nutrition, 75, 100

O

OH, 110, 111, 112, 113, 114, 116, 117, 118, 119, 120, 150, 156
operon, 5, 6, 7, 8, 10, 11, 43, 44, 86, 127, 128, 140
opportunistic infections, viii, 29, 32
oral cavity, vii, xi, 2, 30, 137, 138, 144, 158
oral cavity microbiome, vii, xi, 137
organ, 41
organic compounds, 150
organic matter, 150
organism, 2, 24, 45, 124, 134, 142, 144, 147
organize, 4
ornithine, 128
osteomyelitis, x, 93, 148
otitis media, 55
outpatients, 59
oxidative stress, 87

P

parasite, 52, 154
pathogenesis, 4, 6, 16, 17, 32, 33, 34, 48, 53, 54, 56, 58, 63, 65, 67, 72, 73, 87, 127, 135, 154, 160, 161, 162
pathogens, vii, ix, xi, 2, 9, 14, 30, 38, 45, 51, 53, 55, 59, 66, 67, 68, 70, 71, 100, 102, 106, 107, 127, 131, 132, 137, 138, 144, 153, 159, 164
pathology, 20
pathways, 70, 143, 145
PCR, 11, 46, 52, 56, 57, 58, 60, 61, 63, 66, 99, 107, 131, 146
penicillin, 35, 40, 53, 55, 56, 58, 95, 141, 142, 143, 145, 161
penis, 76
peptide(s), 5, 6, 7, 10, 40, 42, 71, 73, 75, 86, 87, 139, 140, 144, 145, 150, 155
peptide chain, 40
periodontal, 113, 139, 147, 149, 154, 158, 160, 162
periodontal disease, 160
periodontitis, 110, 120, 139, 146, 147, 148, 158, 159, 162, 164
peritonitis, 32, 33, 34, 55, 100, 104, 140
permeability, 35, 95, 151
pH, 6, 30, 110, 112, 113, 118, 119, 124, 126, 133, 147, 149, 151, 152, 158
phage, 78, 144
phagocyte, 34
phagocytic cells, 75
phagocytosis, ix, 5, 8, 15, 64, 69, 74, 149, 160, 163
pharmaceutical, 21
pharmacotherapy, 20, 26
phenol, 119
phenotype(s), ix, 9, 16, 20, 30, 35, 37, 39, 41, 42, 43, 44, 49, 56, 62, 64, 67, 73, 78, 90, 129, 153, 161, 163, 164
Pheromones, v, 5, 69
Philadelphia, 106
phospholipids, 71
phosphorous, 118
phosphorylation, 36
physicians, 34
physiology, 17
pigs, 48, 51, 97
pilot study, 60
pipeline, 21
plants, 20, 21, 22, 24, 30, 94, 124
plasmid, 4, 5, 6, 8, 10, 11, 12, 16, 35, 37, 39, 42, 44, 54, 58, 62, 66, 74, 75, 78, 79, 87, 89, 90, 96, 98, 105, 154
plasticity, 47, 83
platelets, 86
PM, 86, 106
pneumonia, 40, 66
point mutation, 44
polar, 142
polymer, 6, 118
polymerase, 46, 66, 85, 146, 153, 160
polymerase chain reaction, 46, 66, 85, 146, 153, 160
polymers, 84
polymorphism, 46, 51, 149
polypeptide(s), 42, 140
polysaccharide(s), ix, 3, 5, 8, 12, 17, 33, 69, 71, 73, 74, 85, 86, 140, 156, 163
population, 8, 57, 73, 83, 91, 97, 100, 101, 105, 121, 156, 160, 162
population density, 8, 73
population structure, 83, 91, 101, 105
Portugal, 1, 10, 19, 60, 90
potassium, 39, 118, 152
poultry, 37, 48
precipitation, 118
preparation, xi, 110, 123, 124, 150, 162
preservation, 132, 135
prevention, 45, 81, 130, 143
priming, 71
principles, 85, 149
probe, 52
probiotic(s), vii, ix, xi, 20, 30, 69, 70, 83, 85, 94, 123, 124, 138
producers, x, 47, 73, 93, 99, 126
pro-inflammatory, ix, 69, 70, 71, 74

proliferation, 113
promoter, 37, 41, 97
propagation, 7
prophylactic, 13, 97, 157
prophylaxis, x, 56, 94, 103
propylene, x, 109, 111, 112
protection, 8, 38, 81, 162
protein synthesis, 35, 37, 38, 143, 156
proteinase, 32, 126, 133
proteins, 6, 7, 8, 12, 13, 16, 20, 32, 33, 34, 35, 37, 38, 40, 43, 44, 70, 73, 84, 88, 95, 110, 124, 128, 139, 140, 141, 150, 154, 159, 161, 163
proteome, 83, 84
protons, 119
pseudogene, 33
Pseudomonas aeruginosa, 61
pulp, 118, 147, 159
pulp tissue, 159
pumps, 40, 81

Q

Queensland, 69

R

radiation, 90
radicals, 11
RE, 105
reactions, 46
reactive oxygen, 6, 34, 71, 87, 139
reading, 5, 7, 15, 43
real time, 46
receptors, 71, 75
recognition, ix, 47, 69
recombination, 39, 78, 91, 106
recommendations, 143
recovery, 41, 155
recurrence, 45
red blood cells, 32, 128
regions of the world, ix, 30
rehabilitation, 148
renal failure, x, 41, 94, 103
replication, 38, 140
researchers, 26, 138
residues, 159
resolution, x, 70, 93, 102
resources, 149
response, ix, 5, 8, 13, 50, 57, 61, 69, 70, 71, 86, 87, 89, 90, 94, 110, 113, 141, 144, 147, 149, 157
restoration, 148
restriction enzyme, 46

retail, 15, 92
retirement, 121
ribosome, 37, 38, 95, 129, 143
risk(s), vii, x, 1, 12, 31, 39, 41, 45, 53, 57, 67, 91, 93, 101, 102, 103, 107, 143
risk factors, 12, 31, 39, 41, 53, 101, 102
rods, 150
room temperature, 111, 112
root(s), vii, x, xi, 7, 32, 109, 110, 111, 112, 113, 116, 117, 118, 119, 120, 121, 137, 138, 139, 144, 146, 147, 148, 149, 150, 151, 152, 153, 154, 155, 156, 158, 159, 160, 162, 163, 164
root canal system, x, 109, 110, 119, 146, 147, 148, 149, 150, 151
root canal treatment, vii, xi, 110, 111, 117, 121, 137, 146, 153, 163
routes, 147

S

safety, 91, 92, 134
salinity, 124, 130, 147
saliva, 139, 162, 164
Salmonella, 103
salt concentration, 20, 30, 71
salts, 34, 151
scanning electron microscopy, xi, 109
scarcity, 43
science, 89
secondary metabolism, 21
secrete, 4
secretion, vii, 1, 7, 10, 70, 73, 139
selectivity, 27
senses, 50, 87, 144
sensing, 6, 8, 14, 15, 32, 55, 73, 85, 139, 144, 145, 154, 156, 160, 161, 163
sensitivity, 46, 100
sensors, 157
sepsis, 2, 84, 100
septic arthritis, 31
septum, 143
sequencing, 43, 47, 82
Serbia, xi, 123, 125, 132, 133
serine, 6, 7, 15, 32, 42, 43, 139, 142, 144, 154, 160
serum, 5, 34, 38, 149, 155
sewage, 99
sex, vii, 1, 4, 5, 10, 11, 83, 87, 89, 90, 128, 154, 156
sheep, 32, 97, 98
shock, 89
showing, 39, 40, 44, 100, 101, 127
signalling, 70, 73, 78
signals, 73, 88
signs, vii, ix, 69

skeleton, 27
skin, 2, 7, 30, 39, 40
SNP, 88
social behavior, 83
sodium, 120, 147, 149, 150, 152, 153, 154, 155, 158, 160, 162, 164
solubility, 112, 113, 118, 151
solution, xi, 109, 111, 112, 113, 114, 119, 149, 159
South Africa, 102
South America, 56
Spain, 49, 58, 66, 132
spore, 149
Spring, 83
SS, 84, 91, 98, 106
stability, 72
stabilization, 73
standardization, 112
staphylococci, 30, 31, 32, 40, 52, 63, 68
starvation, 71, 84, 147, 149, 158
state(s), xi, 8, 47, 62, 71, 84, 123, 124, 147, 149
sterile, vii, ix, 69, 98, 100, 113, 114, 148, 150
streptococci, 6, 13, 14, 36, 54, 86, 134, 141
stress, 34, 70, 71, 84, 156
stress response, 34, 71, 84
structure, 12, 21, 24, 51, 54, 88, 106, 132, 161
substitution, 42
substrate(s), 6, 10, 14, 36, 128
success rate, 149
sulfate, 140
Sun, 57, 161
surface component, 33, 75, 140, 144, 159
surface structure, 89
surrogates, 85
surveillance, 45, 46, 50, 52, 56, 57, 61, 62, 63, 103
survival, xii, 32, 33, 64, 70, 71, 74, 83, 87, 100, 110, 119, 137, 139, 140, 146, 147, 149, 155, 157, 163
susceptibility, 32, 37, 39, 40, 43, 44, 48, 49, 50, 53, 54, 55, 56, 58, 59, 61, 63, 65, 74, 90, 91, 92, 103, 105, 132, 140, 141, 151, 154, 158, 159, 160
Sweden, 41, 63
swelling, 148
symbiosis, 26
symptoms, vii, ix, 69
synergistic effect, 35, 36, 142
synovial fluid, 6
synthesis, xi, 5, 10, 24, 27, 37, 40, 42, 43, 44, 62, 95, 123, 125, 141, 142, 144

T

Taiwan, 44
target, 7, 21, 36, 37, 38, 43, 50, 87, 113, 140, 143
taxonomy, 16, 53, 65

techniques, xii, 21, 46, 67, 120, 137, 146, 147, 148, 153
technology, 133
teeth, xi, 109, 110, 113, 116, 117, 118, 119, 120, 146, 147, 152, 158, 159, 160, 163, 164
teicoplanin, 9, 42, 43, 44, 57, 81, 97
temperature, 30, 71, 124
testing, 46, 50, 53, 63, 103, 141, 150, 152
tetracyclines, viii, 9, 30, 34, 37, 38, 66, 79
texture, 126
therapeutic agents, 81
therapeutic use, x, 93
therapy, ix, x, 12, 30, 31, 39, 44, 58, 59, 60, 66, 109, 110, 113, 118, 119, 138, 140, 142, 146, 153, 155, 156, 163
third-generation cephalosporin, viii, 29, 102
threonine, 142
thrombus, 57
time periods, 45
tissue, vii, 1, 6, 30, 40, 75, 94, 100, 110, 112, 113, 119, 145, 147, 148, 150, 159, 162
TLR, 70, 71
TLR2, 84
TNF, 71
TNF-α, 71
tonsils, 57
tooth, x, 109, 110, 147, 148, 150
toxic effect, 164
toxicity, 112
toxin, xii, 12, 32, 54, 89, 137
traits, vii, ix, 1, 9, 24, 30, 31, 34, 35, 40, 47, 70, 72, 77, 78, 79, 81, 83, 89, 97, 138, 160
transcription, 70
transduction, 85, 154
transference, 34, 46
translation, 3, 95, 145
translocation, 16, 32, 33, 94, 99, 100, 158, 160
transmission, ix, 30, 45, 47, 79, 81, 97
transplant, 54, 59
transplantation, 58
transport, 7, 110, 118, 128
transportation, 35
treatment, viii, x, xi, 9, 15, 19, 30, 34, 35, 37, 38, 39, 40, 42, 51, 58, 60, 68, 71, 81, 83, 84, 88, 94, 100, 101, 102, 103, 109, 110, 119, 127, 128, 130, 134, 137, 138, 141, 142, 144, 147, 148, 149, 150, 151, 152, 156, 159, 162, 163
tumorigenesis, 139
Turkey, 50
tyrosine, xi, 124, 128

Index

U

umbilical cord, 6
United, 10, 11, 12, 30, 38, 50, 52, 53, 55, 62, 95, 107
United Kingdom (UK), 38, 55, 107, 111
United States, 10, 11, 12, 30, 50, 52, 55, 62, 95
urban, 129
urban areas, 129
urinary bladder, 3
urinary tract, vii, viii, xi, 1, 2, 3, 9, 16, 19, 29, 30, 31, 33, 48, 49, 53, 55, 57, 58, 63, 85, 86, 87, 100, 101, 104, 127, 137, 138, 139, 140, 141, 158, 161, 162
urinary tract infection, vii, viii, 1, 2, 3, 9, 16, 19, 29, 31, 48, 49, 53, 55, 57, 58, 63, 85, 86, 87, 101, 104, 127, 139, 140, 141, 158, 161, 162
urine, 72, 85
USA, 21, 41, 50, 85, 89, 115
UV, 79

V

vaginal tract, 138
valve, 31, 32, 48, 57
valvular heart disease, 31
Vancomycin-Resistant Enterococci, ix, 17, 30
vector, 45
vegetables, viii, ix, 19, 93
vehicles, x, 109, 110, 111, 113
versatility, viii, 29, 129
virulence factors, vii, x, xi, 1, 2, 9, 10, 17, 32, 49, 71, 73, 74, 77, 78, 84, 85, 92, 93, 94, 97, 104, 105, 110, 127, 133, 137, 138, 139, 144, 153, 155, 163
viruses, 149
viscosity, 112, 151
VRE, ix, 9, 14, 21, 22, 28, 30, 37, 38, 40, 41, 42, 44, 45, 46, 47, 64, 102, 144

W

Wales, 57
war, 45, 60
Washington, 11, 14, 48, 53, 54, 55, 56, 58, 64, 65, 67, 134, 135
water, vii, viii, xi, 1, 2, 9, 19, 30, 64, 99, 109, 111, 118, 124, 152, 159
Western Europe, 63
white blood cells, 32
wool, 114
workers, 45, 138
worldwide, 2, 35, 47, 48, 138
wound infection, viii, 2, 29, 31

X

xanthones, 21

Y

yeast, 121, 133, 150
yield, 39

Z

zinc, 139, 152